Environmental Science and Engineering

Series Editors

The ultimate goal of this series is to contribute to the protection of our environment, which calls for both profound research and the ongoing development of solutions and measurements by experts in the field. Accordingly, the series promotes not only a deeper understanding of environmental processes and the evaluation of management strategies, but also design and technology aimed at improving environmental quality. Books focusing on the former are published in the subseries Environmental Science, those focusing on the latter in the subseries Environmental Engineering.

Gordon Huang · Ed McBean · Yongping Li ·
Wendy Huang · Peng Zhang
Editors

15th International Conference on Environmental Science and Development (ICESD2024)

Springer

Editors
Gordon Huang
Faculty of Engineering and Applied Science
University of Regina
Regina, SK, Canada

Ed McBean
School of Engineering
University of Guelph
Guelph, ON, Canada

Yongping Li
School of Environment
Beijing Normal University
Beijing, China

Wendy Huang
Department of Civil Engineering
University of Calgary
Calgary, AB, Canada

Peng Zhang
Institute for Energy, Environment
and Sustainable Communities
Regina, SK, Canada

ISSN 1863-5520 ISSN 1863-5539 (electronic)
Environmental Science and Engineering
ISBN 978-3-031-88682-9 ISBN 978-3-031-88683-6 (eBook)
https://doi.org/10.1007/978-3-031-88683-6

This Springer imprint is published by the registered company Springer Nature Switzerland AG
The registered company address is: Gewerbestrasse 11, 6330 Cham, Switzerland

Preface

2024 15th International Conference on Environmental Science and Development (ICESD 2024) was held successfully during August 7–9, 2024 in Calgary, Canada. It aims to provide a unique platform for researchers, scientists, engineers, and professionals from all over the world to present their latest research results and new ideas in terms of Environmental Science and Development. We deeply appreciate their active participation in ICESD 2024.

ICESD 2024 is supported by the International Society for Environmental Information Sciences. The conference received strong support from experts, practitioners, and researchers from various countries. During the conference, Dr. Yongping Li from Beijing Normal University, China, and Dr. Wendy Huang from the University of Calgary, Canada, provided wonderful keynote presentations.

ICESD 2024 has received submissions contributed by researchers from many countries such as Canada, Japan, China, the UK, Latvia, South Korea, Thailand, India, Pakistan, Uganda, Peru, Hungary, Egypt, Philippines, Malaysia, Indonesia, Bangladesh, Saudi Arabia, Russia, and Colombia. After two rounds of rigorous peer reviews, 13 papers were selected in this proceeding, within the scope of Environmental Science and Development such as solid waste management, pollution remediation, resource management, biodiversity conservation, energy-system optimization, carbon-emission reduction, green technology, and sustainable development. Herewith, we'd like to sincerely appreciate the technical committee members for reviewing the submitted papers.

Regina, Canada
November 2024

Gordon Huang
Honorary Chair

Committees

Honorary Chair

Gordon Huang	University of Regina, Canada

Conference Chairs

Edward McBean	University of Guelph, Canada
Yongping Li	Beijing Normal University, China
Wendy Huang	University of Calgary, Canada

Publicity Chairs

Amrane Abdeltif	University of Rennes1, France
Zhen Hu	Shandong University, China

International Programme Committee Members

Guangwei Huang	Sophia University, Japan (Co-chair)
Changping Chen	Xiamen Ocean Vocational College, China (Co-chair)
Amrane Abdeltif	University of Rennes1, France
Sajjad Ahmad	University of Nevada, USA
Brian Baetz	McMaster University, Canada
Valentina Busini	Politecnico di Milano, Italy
Lei Chen	Beijing Normal University, China
Xi Chen	Tianjin University, China
George Christakos	San Diego State University, USA
Thomas Donnelly	University of Newcastle, Australia
Subhasis Ghoshal	McGill University, Canada

Oliver Günther	University of Potsdam, Germany
Qiang Huang	Xi'an University of Technology, China
Younggy Kim	McMaster University, Canada
Daniel Loucks	Cornell University, USA
Guangheng Ni	Tsinghua University, China
Jinren Ni	Peking University, China
Wei Ouyang	Beijing Normal University, China
Jeffrey Romm	University of California at Berkeley, USA
Zhenyao Shen	Beijing Normal University, China
William Sonzogni	University of Wisconsin-Madison, USA
Eric Strauss	Michigan State University, USA
Masayuki Tamura	Kyoto University, Japan
Fuqiang Tian	Tsinghua University, China
Caterina Valeo	University of Victoria, Canada
Shuguang Wang	Shandong University, China
Xinhao Wang	University of Cincinnati, USA
Huining Xiao	University of New Brunswick, Canada
Lihua Xiong	Wuhan University, China
Julian Yeomans	York University, Canada
Yujun Yi	Beijing Normal University, China
Guangming Zeng	Hunan University, China

Organizing Committees

Peng Zhang	University of Regina, Canada (Co-chair)
Yao Yao	University of Regina, Canada (Co-chair)
Wendy Huang	University of Calgary, Canada (Co-chair)
Luoyilang Ke	University of Calgary, Canada
Yanfeng Li	Beijing Normal University, China
Mengsi Liu	University of Calgary, Canada
Bin Luo	University of Regina, Canada
Shuang Qiao	University of Calgary, Canada
Tangnyu Song	University of Regina, Canada
Feng Wang	Beijing Normal University, China
Xiong Zhou	Beijing Normal University, China

Technical Committees

Fenty Puluhulawa	Gorontalo State University, Indonesia
Tank Shantilal K.	V. N. South Gujarat University, India
Violeta-Carolina Niculescu	National Research and Development Institute for Cryogenics and Isotopic Technologies-ICSI Ramnicu Valcea, Romania
Dong-Sheng Jeng	Griffith University School of Engineering and Built Environment, Australia
Ahmad Zuhairi bin Abdullah	Universitiy Sains, Malaysia
Hameed Sulaiman	Sultan Qaboos University, Oman
Shabir Hussain Khahro	Prince Sultan University, Riyadh, Saudi Arabia
Sivakumar Nallusamy	Sultan Qaboos University, Oman
Lee Hwang Sheng	Universiti Tunku Abdul Rahman, Malaysia
Jenny Ong	SEGi University, Malaysia
Desh Raj Thakur	H. P. University, India
Huang Wei-Po	National Taiwan Ocean University, Taiwan
Low Sheau-Ting	Universiti Teknologi Malaysia, Malaysia
Eric James Strauss	Michigan State University, USA
Mohammed Muayad T. A.	Environment Research Center, University of Technology, Iraq
Ivana Plazonić	University of Zagreb, Faculty of Graphic Arts, Croatia
Suresh Prasad Singh	Jharkhand Rai University, India
Ramesh A/L T. Subramaniam	Universiti Malaya, Malaysia
Yu-Long Chao	National Formosa University, Taiwan
Sajjad Ahmad	University of Nevada, USA
Tanakorn Wongwuttanasatian	Khon Kaen University, Thailand
Jorge Ferrer	Universidad Continental, Perú
Halil Ibrahim Burgan	Akdeniz University, Türkiye
Suchit Kumar Patel	Central University of Jharkhand, India
Zhen Hu	Shandong University, China
Xiaosheng Qin	Nanyang Technological University, Singapore
Voon Loong Wong	Heriot-Watt University Malaysia, Malaysia
Srinath Suranani	National Institute of Technology Warangal, India
Amrane Abdeltif	University of Rennes1, France
Anjana P. Anantharaman	National Institute of Technology Warangal, India
Kinga Szopińska	Bydgoszcz University of Science and Technology, Poland
Ng Jo-Han	University of Southampton Malaysia, UK
Zhongsheng Guo	Institute of Soil and Water Conservation, Northwest A&F University, China

Fauziah Shahul Hamid	University of Malaya, Malaysia
Jigna Desai	Veer narmad South Gujarat University, India
Amir Abdul Razak	Universiti Malaysia Pahang, Malaysia
Murtada D. Naser	University of Basrah, Iraq
Amaal Gh. Yasser	University of Basrah, Iraq
Ahmed Saqr	Mansoura University, Mansoura, Egypt
Baba Imoro Musah	Xishuangbanna Tropical Botanical Garden Chinese Academy of Sciences (XTBG-CAS), China
Daniel Adu	Jiangsu University, China
Nina Almasifar	Cukurova University, Türkiye

Contents

Solid Waste Management and Environmental Pollution Remediation

Identifying the Potential Disposal Sites for Effective MSW Disposal in Different Regions of Thailand

Waralak Khongouan(✉)

Faculty of Architecture and Planning, Thammasat University, Pathum Thani, Thailand
waralak.kh@gmail.com

Abstract. Thailand faced a crisis of municipal solid waste during its past. Even though local and national governmental agencies explored the issue of disposing of municipal solid waste, there has never been a comparative, potential analysis of nationwide disposal sites to contribute to the municipal solid waste management. Therefore, the objective of this article is to explore the municipal solid waste disposal situation in Thailand, in terms of municipal solid waste management, the municipal solid waste volume, and the analysis of disposal site potentiality. The data were analyzed by using a scalogram. According to the research results, to have fully equipped disposal sites with technical properties, this can be achieved by either increasing the volume of properly disposed and recycled municipal solid waste, or by decreasing improperly disposed of bulk waste, simultaneously. However, despite the decline in municipal solid waste volume, as well as the better situation of municipal solid waste management in various regions, Thailand still has many inappropriate municipal solid waste disposal sites, which reflects that it is time for Thailand to change to a waste management that emphasizes separating and reusing waste rather than disposing of it to landfill.

Keywords: Disposal Site · Municipal Solid Waste · Scalogram Analysis

1 Introduction

The increasing volume of municipal solid waste has become a significant environmental issue worldwide [1–3] with a global estimated generation rate of 0.74 kilograms per person per day. Accordingly, the global municipal solid waste in 2016 was a total of 2.01 billion tons, while the anticipated volume shall be about 3.10 billion tons in 2050 [4]. Considering the national situation of municipal solid waste, Thailand is confronting a management crisis [5]. By 2023, the total municipal solid waste was approximately 26.95 million tons with 7.47 million tons or 27.72% of the total volume identified as improperly disposed. Besides, Thailand has implemented forms of improper disposal such as controlled dumps, open dumps, open burns, etc. accounting for up to 1,964 sites or 94.42% of the overall municipal solid waste disposal sites in Thailand [6]. The increasing volume, plus the improper disposal system, certainly creates problems afterwards, whether it is an unpleasant odor, greenhouse gas emission, water pollution,

G. Huang et al. (Eds.): ICESD 2024, ESE, pp. 3–16, 2025.
https://doi.org/10.1007/978-3-031-88683-6_1

or poor hygienic situations, etc. [7]. Moreover, providing an area for a disposal site causes public conflict and reluctance as well [8, 9].

In the past, local administrative organizations significantly took responsibility for collecting, transporting, and disposing of generated municipal solid waste in Thailand, under the national authorization of the Waste and Hazardous Substance Management Division, Pollution Control Department. Although this agency has explored and studied the area characteristics and the problems from each disposal site all over the country, there is still no complete data comparative analysis able to be used to acknowledge the potentiality and problem of regional and national disposal sites as a matter of fact. Therefore, the objective of this article is to study the municipal solid waste disposal situation in Thailand, in terms of municipal solid waste management, the municipal solid waste volume, and the analysis of disposal site potentiality. As a result, the derived data then contributes to the relevant local and national agencies for solving inefficient municipal solid waste management, reliably handling the municipal solid waste situation in the future.

2 Literature Review

The considered methods of disposing of municipal solid waste are different, namely by landfill, compost, incineration, and open dumping, etc. Therefore, the resolved method depends on the identified category and volume of municipal solid waste, the area size of disposal site, and the implementation budget [7, 10]. In the case of Thailand, the governmental sector has promoted landfilling and incineration. The volumes and corresponding area sizes required are as follows: a volume smaller than 15 tons per day needs an area size of at least 16,000 m^2; a volume of 15–50 tons per day needs at least 32,000 m^2; a volume of 51–100 tons per day needs at least 112,000 m^2; a volume of 101–300 tons per day needs at least 208,000 m^2; a volume of 301–700 tons per day needs at least 384,000 m^2; and a volume of 700 tons up per day needs at least 448,000 m^2 [11]. In the disposal sites for landfilling and open dumping, it is essential to be equipped with a wastewater treatment system, a weighing scale, a bulldozer, and a machine-washing area. Moreover, the area size must suit municipal solid waste bulk and the treatment approach, as well as a service-opening duration long enough to generate the accumulated municipal solid waste [8].

However, the major factor in achieving successful municipal solid waste management is not to dispose of the municipal solid waste by landfill, but instead pragmatically sorting waste at the origin, purposely recycling and reducing municipal solid waste. It is essential to strengthen public awareness and the realization of authorized personnel to achieve this practical procedure [7, 12, 13].

3 Methodology

The study of municipal solid waste disposal situation in Thailand were composed of (1) studying the municipal solid waste management, (2) studying the municipal solid waste volume, and (3) analyzing the disposal site potentiality. For the study of municipal solid waste management involves collecting and studying secondary data, both quantitative and qualitative, regarding management plan and policy, authorized agency, and the municipal solid waste disposal site. The updated data in April, 2024, which was collected by the Waste and Hazardous Substance Management Division, Pollution Control Department, the authorized national bureau [6].

The study of municipal solid waste volume is both a collection of quantitative secondary data, the municipal solid waste volume of Thailand between 2014 and 2023, the updated data in April, 2024, which was collected by the Waste and Hazardous Substance Management Division, Pollution Control Department, and an analysis of change in the municipal solid waste volume data includes the generated volume, the proper-disposed volume, the improper-disposed volume, and the recycled volume. Consequently, the mentioned data is illustrated by the 6 regional boundaries of Thailand, namely the Northern (9 provinces), the Central (22 provinces), the Northeastern (20 provinces), the Western (5 provinces), the Eastern (7 provinces), and the Southern (14 provinces) regions [6].

An analysis of disposal site potentiality involves collecting quantitative and qualitative secondary data on municipal solid waste disposal sites that are open for service and methods for disposing of municipal solid waste by landfill or open dump (the system of landfilling and open dumps are mainly operated in Thailand), the updated data in April, 2024, which was collected by the Waste and Hazardous Substance Management Division, Pollution Control Department. The disposal site data has a total of 88 items consisting of 10 sites in the Northern region, 24 sites in the Central, 31 sites in the Northeastern, 9 sites in the Eastern, 14 sites in the Southern, and 2 sites in the Western (see Fig. 1) [14]. Additionally, identification of the disposal site potentiality through systematized data with Scalogram analysis was developed by Louis Guttman [15]. The significant properties of landfilling and open dump sites for Scalogram analysis include as follows: the service-opening period of less than 20 years, at least 10 local administrative organizations using the service, landfilling as the disposal system, more than 50 tons of input municipal solid waste per day, a disposal area size suiting the operated system and the volume, the installation of a wastewater treatment system, the installation of a weighing scale, an available bulldozer, and a provided machine-washing area [8]. Furthermore, the potentiality classification or hierarchy of municipal solid waste disposal sites, separates into 4 groups, namely high-potential, moderate-potential, low-potential, and no potential, followed by order of score disperse (2 sites in the Western, which cannot be used to analyze the potential of municipal solid waste disposal sites with the Scalogram analysis).

4 Findings and Results

4.1 A Study of Municipal Solid Waste Management

The municipal solid waste management in Thailand, is carried out by the local administrative organizations which is the smallest governance units in Thailand, and usually emphasizes the elimination of large quantities of municipal solid waste through various methods more than reducing and sorting municipal solid waste.

The 10 disposal sites in the Northern

1 = Uttaradit Town Municipality
2 = Chiang Mai Provincial Administrative Organization
3 = Pamat Subdistrict Municipality
4 = Chiang Rai City Municipality
5 = Lampang Provincial Administrative Organization
6 = Mae Sai Mittraphap Subdistrict Municipality
7 = Wiang Fang Subdistrict Municipality
8 = Lampang City Municipality
9 = Nan Town Municipality
10 = Bandu Subdistrict Municipality

The 24 disposal sites in the Central

1 = Hankha Subdistrict Municipality
2 = Nakhonsawan City Municipality
3 = Nakhonpathom City Municipality
4 = Kamphaeng Phet Town Municipality
5 = Phra Nakhon Sri Ayutthaya Provincial Administrative Organization
6 = Bangmulnak Town Municipality
7 = Uthaithani Town Municipality
8 = Takhli Town Municipality
9 = Phetchabun Town Municipality
10 = SuphanBuri Town Municipality
11 = Chumsaeng Town Municipality
12 = Nongphai Subdistrict Municipality
13 = Angthong Town Municipality
14 = Nakhonloung Subdistrict Municipality
15 = Taphanhin Town Municipality
16 = Songpeenong Town Municipality
17 = Sriprachan Subdistrict Municipality
18 = Banmi Town Municipality
19 = Nongkhae Subdistrict Municipality
20 = Lomsak Town Municipality
21 = Salokbat Subdistrict Administrative Organization
22 = Khoksumrong Subdistrict Municipality
23 = Promburi Town Municipality
24 = Huai Phlu Subdistrict Municipality

The 9 disposal sites in the Eastern

1 = Rayong Provincial Administrative Organization
2 = Kanghangmaeo Subdistrict Administrative Organization
3 = Muang Klaeng Subdistrict Municipality
4 = Prachin Buri Town Municipality
5 = Chanthaburi Town Municipality
6 = Nongprue Town Municipality
7 = Bangkhla Subdistrict Municipality
8 = Kabin Subdistrict Municipality
9 = Santung Subdistrict Municipality

The 31 disposal sites in the Northeastern

1 = Kalasin Town Municipality
2 = Udon Thani City Municipality
3 = Warinchamrab Town Municipality
4 = Muangphon Town Municipality
5 = Ban Phai Town Municipality
6 = Na Klang Subdistrict Municipality
7 = Non Sang Subdistrict Municipality
8 = Buriram Town Municipality
9 = Yasothon Town Municipality
10 = Roi Et Town Municipality
11 = Siraket Town Municipality
12 = Nakhon Ratchasima City Municipality
13 = Mukdahan Town Municipality
14 = Bandung Town Municipality
15 = Sikhoraphum Subdistrict Municipality
16 = Phibun Mangsahan Town Municipality
17 = Kranuan Town Municipality
18 = Nong Bua Lamphu Town Municipality
19 = Sikhio Town Municipality
20 = Sungnoen Subdistrict Municipality
21 =Kaikhum Subdistrict Municipality
22 = Loei Town Municipality
23 =Satuk Subdistrict Municipality
24 = Chiagyuen Subdistrict Municipality
25 = Maha Sarakham Town Municipality
26 = Chaiyaphum Town Municipality
27 = Muangpak Town Municipality
28 =Kamphaeng Subdistrict Municipality
29 =Banphue Subdistrict Municipality
30 =Banborabue Subdistrict Municipality
31 = Wapeepathum Subdistrict Municipality

The 14 disposal sites in the Southern.

1 = Kun Trang Town Municipality
2 = Kamphaeng Subdistrict Municipality
3 = Satun Town Municipality
4 = Trang City Municipality
5 = Yala City Municipality
6 = Sungai Kolok Town Municipality
7 = Betong Town Municipality
8 = Phangnga Town Municipality
9 = Nakhon Si Thammarat City Municipality
10 = Banpru Town Municipality
11 = Pakphanang Town Municipality
12 = Chumphon Town Municipality
13 = Dondak Town Municipality
14 = A.C.J. Construction Limited Partnership

Fig. 1. The disposal sites for the Scalogram analysis.

The municipal solid waste management in Thailand, performs under the National Waste Management Action Plan No. 2 (2022–2027) [16], purposely promoting the clustering of area management in order to reduce the budget and to promote proper management and suitability. The local administrative organizations, therefore have assembled to manipulate the municipal solid waste altogether.

Therefore, the management clusters are categorized into the following areas: a large size area (Model L) with the carrying capacity of more than 300 tons per day; a medium size area (Model M) with the carrying capacity of 50–300 tons per day; and a small size area (Model S) with the carrying capacity of less than 50 tons per day. Each local administrative organization either by itself or by hiring from the private sector, collects the generated municipal solid waste, then conveys towards the disposal site within its cluster. In addition, the disposal site in each management cluster properties differently. According to the data of the Waste and Hazardous Substance Management Division, Pollution Control Department, [6] Thailand in 2023 still had considerable numbers of municipal solid waste disposal sites operating with improper procedures, such as controlled dumping, open dumping, open burning, etc. These account for up to 1,964 sites or 94.42% of the municipal solid waste disposal sites in Thailand. On the other hand, there are only 116 proper disposal sites, such as landfills, incineration-waste to energy (WTE), refuse derived fuel (RDF), etc. These account for 5.58% of the municipal solid waste disposal sites in Thailand. Nevertheless, the number of municipal solid waste disposal sites in Thailand was diminishing during 2014–2023 since building a municipal solid waste disposal site is often opposed by local communities [8, 17].

4.2 A Study of Municipal Solid Waste Volume

The procedure of sorting waste with more proper, substantial management of municipal solid waste during 2014–2023. As a result, all regions have more properly disposed municipal solid waste and more recycled municipal solid waste; simultaneously, improperly disposed municipal solid waste decreased [6]. However, Thailand increased the amount of municipal solid waste in 2023. The entire volume of generated municipal solid waste was approximately 26.95 million tons, or 0.76 million tons more than 26.19 million tons in 2014 with a rate of change increasing by 0.32% annually. In the case of classifying the aforementioned municipal solid waste volume by the regional administrative boundaries of Thailand, it reveals that the Northern, the Northeastern, and the Western regions had a decreasing volume during 2014–2023, meanwhile the municipal solid waste volume in the Central, the Eastern, and the Southern regions increased. The region with the highest amount of municipal solid waste in Thailand throughout the past period was the Central, which generated up to 11.44 million tons in 2023 [6]. This is because the region has the most provinces (22) whilst also being the location of Bangkok Metropolis, the capital of Thailand (see Figs. 2 and 3).

4.3 An Analysis of Disposal Site Potentiality

According to analyzing the potential of disposal sites, it was found the municipal solid waste disposal sites are mostly in the moderate to low potential groups in all regions. The potentiality in the Central, the Northeastern, and the Eastern regions is classified into 4 groups, namely high-potential, moderate-potential, low-potential, and no potential. Meanwhile, the potentiality in the Northern and the Southern regions is classified into 3 groups, namely high-potential, moderate-potential, and low-potential. The disposal sites with high potential in all regions have all significant properties, which belong to landfill sites and open dump site, otherwise they lack only a small number of properties. Only two municipal solid waste disposal sites have every single significant property, namely

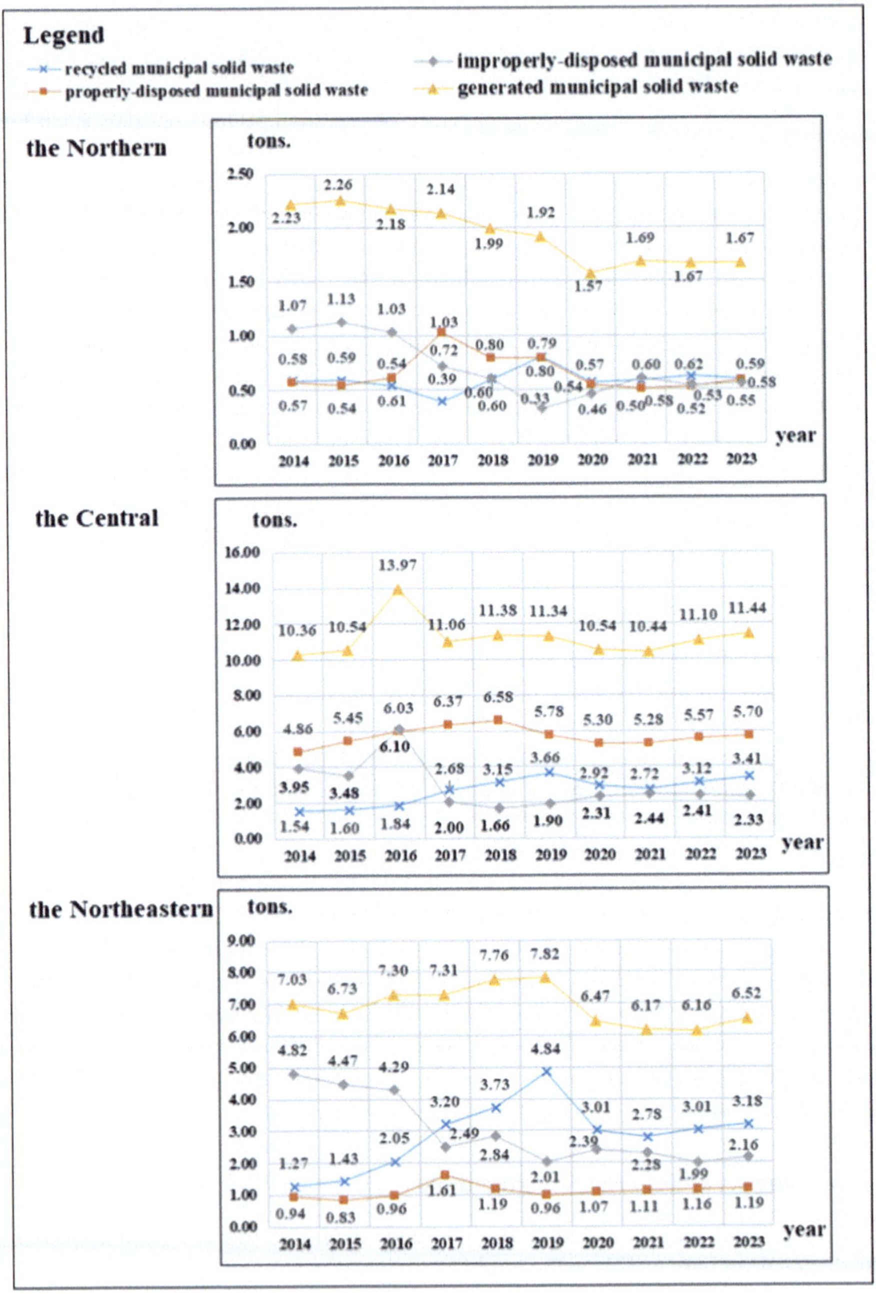

Fig. 2. The municipal solid waste volume in Thailand's northern, central, and northeastern regions.

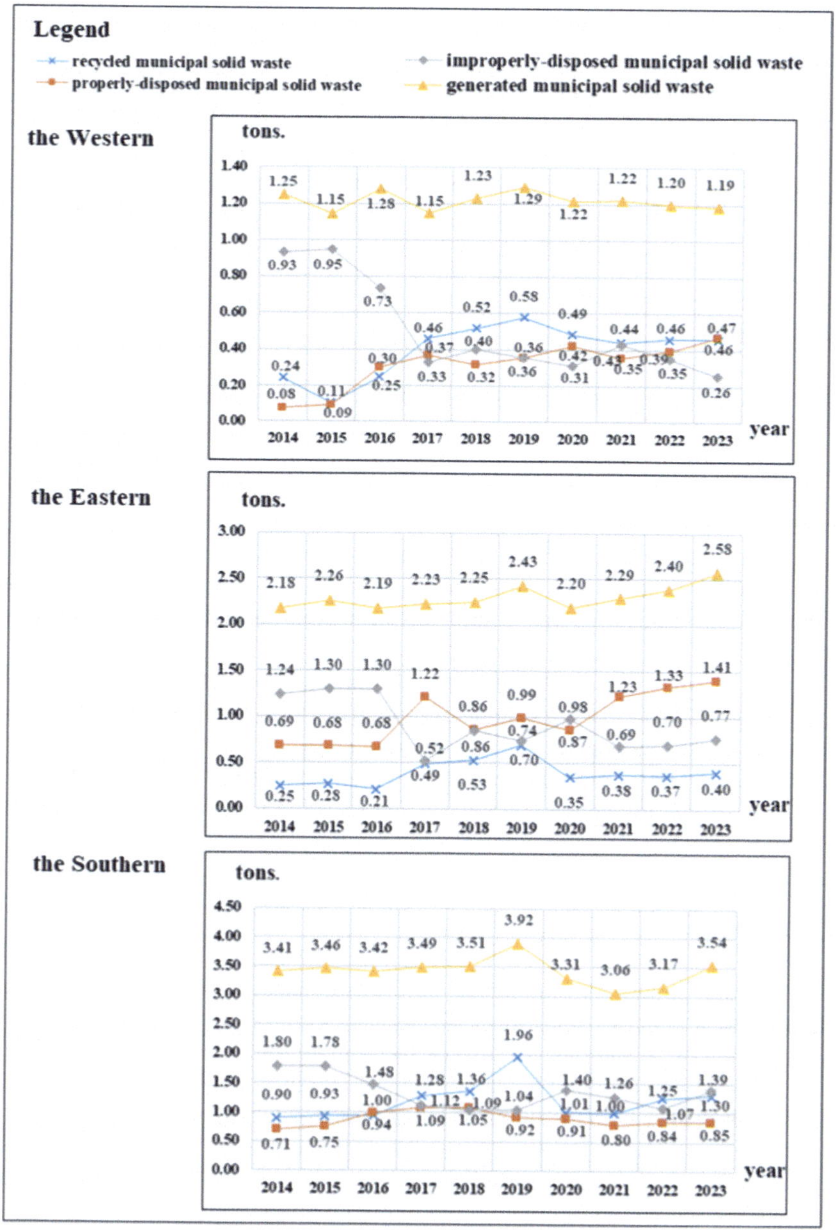

Fig. 3. The municipal solid waste volume in Thailand's western, eastern, and southern regions.

the disposal site of the Hankha Subdistrict Municipality, Chainat province in the Central region, and the disposal site of the Rayong Provincial Administrative Organization, Rayong province in the Eastern region. In the case of the latter, it is located in the province where both a nationally-important industrial estate and port are as well (see Figs. 4, 5, 6 and 7.)

As among the no potential group, these municipal solid waste disposal sites do not have any significant properties of the landfill and open dump sites, namely the disposal site of the Huai Phlu Subdistrict Municipality, Nakhonpathom province in the Central region, the disposal site of the Wapi Pathum Subdistrict Municipality, Maha Sarakham province in the Northeastern region, and the disposal site of the Santung, Trat province in the Eastern region (see Figs. 5, 6, and 7.)

5 Discussion

The municipal solid waste disposal sites in Thailand are still deficient in terms of technical properties. According to the Scalogram analysis results, there are still municipal solid waste disposal sites that have no significant properties to become a landfill site or open dump site. Moreover, the municipal solid waste disposal sites are mostly in the moderate to low potential groups in all regions. This situation is caused by having a limited budget, and lack of area for waste disposal sites. In addition, there is an absence of knowledge and understanding of sustainable waste management that emphasizes reducing and recycling waste rather than completely disposing waste through various methods [18, 19].

However, having completely qualified landfilling and open dump sites is an important factor that causes the increasing volume of properly disposed municipal solid waste and recycled municipal solid waste, as the Scalogram analysis results, there were numerous high-potential disposal sites in the Northeast, in consonance with the increasing volume of properly disposed municipal solid waste and recycled municipal solid waste during 2014–2023.

6 Conclusion

Although a better situation of municipal solid waste management in various regions of Thailand but the municipal solid waste disposal sites are mostly in the moderate to low potential groups in all regions because of incomplete municipal solid waste disposal systems. Meanwhile, Thailand has a continuous increase in the amount of municipal solid waste, altogether with the diminishing of municipal solid waste disposal sites, is a reminder that it is time for Thailand to change its waste disposal model from landfill or open dump to waste separation and waste reuse, such as in countries where successful municipal solid waste management is carried out [20].

The significant properties of disposal sites for scalogram analysis

1 = The service-opening period lesser than 20 years
2 = The at-least 10 local administrative organizations using the service
3 = The landfilling as the disposal system
4 = More than 50 tons of in-put municipal solid waste per day
5 = The disposal area size suiting the operated system and the volume
6 = The installation of wastewater treatment system
7 = The installation of weighing scale
8 = The available bulldozer
9 = The provided machine-washing area

Scalogram analysis process:

1-) filling color in the row of the disposal sites' properties no. 1-9 discovered in each disposal site; or else, unfilled the row, 2) changing the colored row in each column to the proportion from the score of 100, and 3-) total the component scores of each disposal site, then put the rightmost column. Finally, the total scores would then expose each disposal's potential.

Scalogram analysis results: the Northern

Number of the disposal sites	The significant features of disposal sites for scalogram analysis									Total
	4	6	7	3	8	2	5	1	9	
1	11.11	11.11	11.11	12.50		16.67	16.67	16.67	25.00	120.83
2	11.11	11.11	11.11	12.50	14.29	16.67		16.67	25.00	118.45
3	11.11	11.11	11.11		14.29	16.67		16.67	25.00	105.95
4	11.11	11.11	11.11	12.50	14.29		16.67		25.00	101.79
5	11.11	11.11	11.11	12.50		16.67	16.67	16.67		95.83
6	11.11	11.11	11.11	12.50	14.29		16.67	16.67		93.45
7	11.11	11.11	11.11	12.50	14.29		16.67	16.67		93.45
8	11.11	11.11	11.11	12.50	14.29		16.67			76.79
9	11.11	11.11	11.11	12.50	14.29	16.67				76.79
10						16.67				16.67

Scalogram analysis results: the Southern

Number of the disposal sites	The significant features of disposal sites for scalogram analysis									Total
	6	7	8	1	3	4	2	5	9	
1	7.69	7.69	7.69	8.33	11.11	11.11	12.50	20.00		86.13
2	7.69	7.69	7.69	8.33	11.11	11.11	12.50	20.00		86.13
3	7.69	7.69	7.69	8.33	11.11		12.50	20.00		75.02
4	7.69	7.69	7.69	8.33	11.11	11.11	12.50			66.13
5	7.69	7.69	7.69	8.33	11.11	11.11	12.50			66.13
6	7.69	7.69	7.69	8.33	11.11			20.00		62.52
7	7.69	7.69	7.69	8.33	11.11			20.00		62.52
8		7.69	7.69	8.33		11.11	12.50			47.33
9	7.69	7.69	7.69			11.11	12.50			46.69
10	7.69	7.69	7.69		11.11	11.11				45.30
11	7.69	7.69	7.69	8.33			12.50			43.91
12	7.69	7.69	7.69			11.11				34.19
13	7.69	7.69	7.69	8.33						31.41
14	7.69			8.33		11.11				27.14

Fig. 4. The scores of disposal site properties in Thailand's northern and southern regions.

Scalogram analysis results: the Central

Number of the disposal sites	The significant features of disposal sites for scalogram analysis									Total
	8	7	6	2	1	4	3	5	9	
1	4.76	5.56	5.56	7.14	7.14	9.09	9.09	10.00	33.33	91.67
2	4.76	5.56	5.56	7.14		9.09	9.09	10.00	33.33	84.53
3	4.76	5.56	5.56	7.14		9.09			33.33	65.44
4	4.76	5.56	5.56	7.14	7.14	9.09	9.09	10.00		58.34
5	4.76	5.56	5.56	7.14	7.14	9.09	9.09	10.00		58.34
6	4.76	5.56	5.56	7.14	7.14	9.09	9.09	10.00		58.34
7	4.76	5.56	5.56	7.14		9.09	9.09	10.00		51.20
8	4.76	5.56	5.56	7.14	7.14		9.09	10.00		49.25
9		5.56	5.56	7.14		9.09	9.09	10.00		46.44
10	4.76	5.56	5.56			9.09	9.09	10.00		44.05
11	4.76	5.56	5.56		7.14		9.09	10.00		42.11
12	4.76	5.56	5.56	7.14	7.14		9.09			39.25
13	4.76	5.56	5.56	7.14		9.09				32.11
14	4.76			7.14	7.14	9.09				28.14
15	4.76	5.56	5.56		7.14					23.02
16	4.76	5.56	5.56		7.14					23.02
17	4.76	5.56	5.56		7.14					23.02
18	4.76	5.56	5.56	7.14						23.02
19	4.76	5.56			7.14					17.46
20	4.76				7.14					11.90
21	4.76			7.14						11.90
22	4.76		5.56							10.32
23					7.14					7.14
24										0.00

Scalogram analysis results: the Eastern

Number of the disposal sites	The significant features of disposal sites for scalogram analysis									Total
	8	6	7	1	2	4	3	9	5	
1	14.29	16.67	16.67	20.00	20.00	25.00	50.00	50.00	100.00	312.62
2	14.29	16.67	16.67	20.00	20.00		50.00			137.62
3	14.29	16.67	16.67	20.00				50.00		117.62
4	14.29	16.67	16.67	20.00	20.00	25.00				112.62
5	14.29	16.67	16.67		20.00	25.00				92.62
6	14.29			20.00		25.00				59.29
7	14.29	16.67	16.67							47.62
8					20.00					20.00
9										0.00

Fig. 5. The scores of disposal site properties in Thailand's central and eastern regions.

Scalogram analysis results: the Northeastern

Number of the disposal sites	The significant features of disposal sites for scalogram analysis									Total
	6	7	8	3	2	5	1	4	9	
1	3.70	3.70	3.70	5.26	5.56		7.14	7.69	9.09	45.86
2	3.70	3.70	3.70	5.26	5.56	6.67		7.69	9.09	45.38
3	3.70	3.70	3.70	5.26	5.56	6.67		7.69	9.09	45.38
4	3.70	3.70	3.70	5.26	5.56	6.67	7.14		9.09	44.83
5	3.70	3.70	3.70	5.26	5.56	6.67	7.14		9.09	44.83
6	3.70	3.70	3.70	5.26	5.56	6.67	7.14		9.09	44.83
7	3.70	3.70	3.70	5.26		6.67	7.14		9.09	39.27
8	3.70	3.70	3.70	5.26	5.56	6.67		7.69		36.29
9	3.70	3.70	3.70	5.26	5.56	6.67		7.69		36.29
10	3.70	3.70	3.70	5.26	5.56	6.67		7.69		36.29
11	3.70	3.70	3.70	5.26	5.56	6.67		7.69		36.29
12	3.70	3.70	3.70		5.56		7.14	7.69		31.50
13	3.70	3.70	3.70		5.56		7.14	7.69		31.50
14	3.70	3.70	3.70	5.26	5.56				9.09	31.02
15	3.70	3.70	3.70	5.26		6.67	7.14			30.18
16	3.70	3.70	3.70	5.26		6.67	7.14			30.18
17	3.70	3.70	3.70	5.26	5.56	6.67				28.60
18	3.70	3.70	3.70				7.14		9.09	27.34
19	3.70	3.70	3.70	5.26					9.09	25.47
20	3.70	3.70	3.70		5.56			7.69		24.36
21	3.70	3.70	3.70		5.56			7.69		24.36
22	3.70	3.70	3.70		5.56			7.69		24.36
23	3.70	3.70	3.70		5.56		7.14			23.81
24	3.70	3.70	3.70	5.26		6.67				23.04
25	3.70	3.70	3.70					7.69		18.80
26	3.70	3.70	3.70				7.14			18.25
27	3.70	3.70	3.70	5.26						16.37
28							7.14		9.09	16.23
29				5.26		6.67				11.93
30							7.14			7.14
31										0.00

Fig. 6. The scores of disposal site properties in Thailand's northeastern regions.

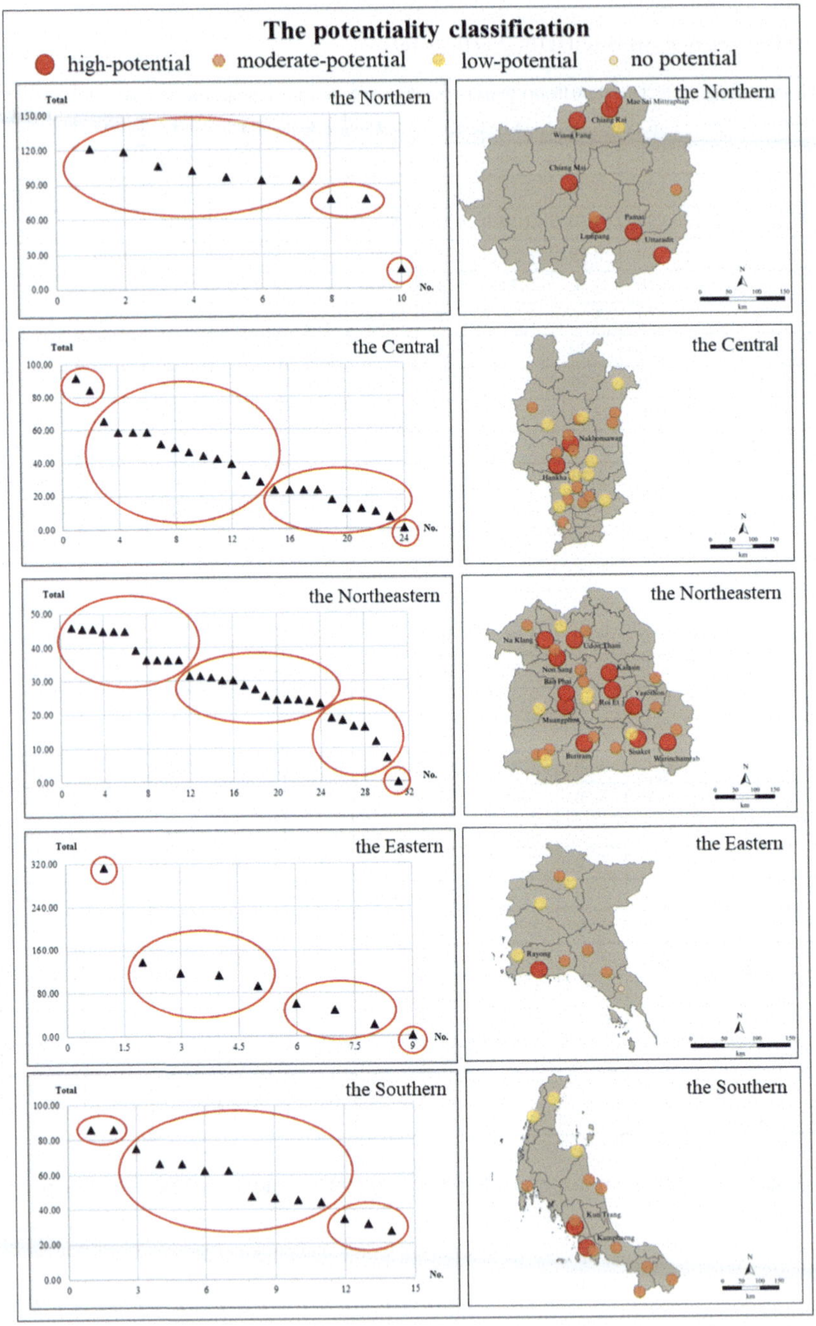

Fig. 7. Scalogram analysis results: Classification of disposal sites according to the potential of disposal sits properties.

References

1. Chen, D.M.C., Bodirsky, B.L., Krueger, T., Mishra, A., Popp, A.: The world's growing municipal solid waste: trends and impacts. Environ. Res. Lett. **15**(7), 1–12 (2020). https://doi.org/10.1088/1748-9326/ab8659
2. Martus, E.: Municipal solid waste management in Russia: Protest, policy, and politics. Russian Anal. Digest **261**, 2–6 (2020). https://doi.org/10.3929/ethz-b-000458206
3. Pheakdey, D.V., Quan, N,V., Khanh, T.D., Xuan, T.D.: Challenges and priorities of municipal solid waste management in Cambodia. Int. J. Environ. Res. Public Health **19**(14), 1–27 (2022). https://doi.org/10.3390/ijerph19148458
4. Kaza, S., Yao, L., Bhada-Tata, P., Woerden, V.F.: What a Waste 2.0: A Global Snapshot of Solid Waste Management to 2050. Washington, DC, World Bank (2018)
5. Eaktasang, N., Suma, Y., Kingkeaw, S.: Feasibility study of municipal solid waste for producing refuse-derived fuel (RDF): a case study of Bangkok, Thailand. IOP Conf. Ser. Earth Environ. Sci. **291**(1), 1–5 (2019). https://doi.org/10.1088/1755-1315/291/1/012010
6. Pollution Control Department.: Information on The Country's Solid Waste Situation (2024). https://thaimsw.pcd.go.th/report_country.php
7. Abubakar, I.R., Maniruzzaman, K.M., Dano, U.L., AlShihri, F.S., AlShammari, M.S., Ahmed, S.M.S., Al-Gehlani, W.A.G., Alrawaf, T.I.: Environmental sustainability impacts of solid waste management practices in the global south. Int. J. Environ. Res. Public Health **19**(19), 1–26 (2022). https://doi.org/10.3390/ijerph191912717
8. Arjwech, R., Somchat, K., Pondthai, P., Everett, M., Schulmeister, M., Saengchomphu, S.: Assessment of geological, hydrogeological and geotechnical characteristics of a proposed waste disposal site: a case study in Khon Kaen, Thailand. Geosciences (Switzerland) **10**(3), 1–17 (2020). https://doi.org/10.3390/geosciences10030109
9. Gutberlet, J., Bramryd, T., Johansson, M.: Expansion of the waste-based commodity frontier: insights from Sweden and Brazil. Sustainability (Switzerland) **12**(7), 1–15 (2020). https://doi.org/10.3390/su12072628
10. Muhammad, L.J., Badi, I., Haruna, A.A., Mohammed, I.A.: Selecting the best municipal solid waste management techniques in Nigeria using multi criteria decision making techniques. Rep. Mech. Eng. **2**(1), 180–189 (2021). https://doi.org/10.31181/rme2001021801b
11. Pollution Control Department.: Announcement of the Pollution Control Department on guidelines for considering and selecting appropriate waste management technology models for local administrative organizations (2020). https://www.pcd.go.th/wp-content/uploads/2020/05/pcdnew-2020-05-25_04-20-32_724764.pdf
12. Sandhu, G.S., Waheed. A.: Waste to wealth: a case study, Islamabad, Pakistan. J. Southwest Jiaotong Univ. **56**(6), 1036–1049 (2021). https://doi.org/10.35741/issn.0258-2724.56.6.90
13. Sandhi, A., Rosenlund, J.: Municipal solid waste management in Scandinavia and key factors for improved waste segregation: a review. Clean. Waste Syst. **8**, 1–10 (2024). https://doi.org/10.1016/j.clwas.2024.100144
14. Pollution Control Department.: The Report on Monitoring of Solid Waste Disposal Sites (2024). https://thaimsw.pcd.go.th/report_pdf.php
15. Peerapun, W.: Analytical Techniques for Regional and Urban Planning, 3rd edn. Charansanitwong Printing Co., Ltd (2013)
16. Pollution Control Department.: The National Waste Management Action Plan No. 2 (2022–2027) (2023). https://www.pcd.go.th/publication/28745
17. Sathabhornwong, S.: Collaborative capacities for successful collaboration: the case of Thai local administrative organizations' waste management. Local Admin. J. **13**(2), 103–136 (2020). https://so04.tci-thaijo.org/index.php/colakkujournals/article/view/240420/166027

18. Alimoradiyan, H., Hajinezhad, A., Yousefi, H., Giampietro, M.: Fostering community participation in sustainable municipal solid waste management at multiple scales in Tehran, Iran. Results Eng. **22**, 102174 (2024). https://doi.org/10.1016/j.rineng.2024.102174
19. Aryal, M., Adhikary, S.: Solid waste management practices and challenges in Besisahar municipality, Nepal. Plos One (2024). https://doi.org/10.1371/journal.pone.0292758
20. Bassey, U., Tom, A., Okono, U., John, M., Sinn, M., Bassey, A., Luke, U., Narra, S.: Characteristics and management of municipal solid waste in Uyo, Akwa Ibom state, Nigeria. Sci. Rep. **14**, 10971 (2024). https://doi.org/10.1038/s41598-024-61108-0

Preliminary Screening of Cadmium Hyperaccumulators

Lanxi Fei[1](✉), Shiqian Xie[2], and Yue Wu[1]

[1] Chengdu Foreign Languages School, Chengdu, China
67371364@qq.com
[2] College of Architecture & Environment, Sichuan University, Chengdu, China

Abstract. With the rapid development of industry and agriculture, soil heavy metal pollution has become increasingly serious, posing a great threat to the environment on which people depend for their survival, and has now become a hot issue of global concern. Phytoremediation has the advantages of economy, environmental protection, easy management and operation, etc. It is recognized as an environmentally friendly in-situ green remediation technology and is widely used in the field of remediation and treatment of heavy metal contaminated soils. The application value of most hyperaccumulators is limited due to their short stature, low biomass, slow growth, long growth cycle and specificity in heavy metal accumulation. Based on this, this paper conducts a screening experiment of potted plants to obtain native hyperaccumulator plants with strong tolerance to cadmium pollution and good cultivation. The plant screening experiment shows that Humulus, Clover, and Fanzha have strong tolerance to cadmium-polluted soil and can grow in soils with high concentrations of cadmium pollution. Humulus has a maximum cadmium content of 56.93 mg/kg and its maximum cadmium accumulation coefficient reaches 1.19, indicating its great potential in the remediation of cadmium-contaminated soils. Clover also shows good accumulation capacity for cadmium, with a maximum cadmium content of 92.52 mg/kg and a maximum accumulation coefficient of 3.87, approaching the critical value for hyperaccumulators. Fanzha has a stronger accumulation capacity, with a maximum cadmium content of 103.03 mg/kg and a maximum accumulation coefficient of 4.08. However, the total biomass of Fanzha is relatively small, making it less suitable for large-scale cultivation and propagation.

Keywords: Soil Pollution · Heavy Metal · Hyperaccumulator · Cadmium

1 Introduction

Phytoremediation technology uses hyperaccumulator to extract, stabilize, or volatilize heavy metals in soils to reduce their toxicity for remediation purposes [1]. The screening of hyperaccumulators remains the most significant and fundamental challenge associated with this technology. Cadmium hyperaccumulators are defined as plants that are capable of absorbing significant quantities of cadmium from the soil, subsequently transferring this element from the roots to the aboveground parts of the plant. In general, there are

G. Huang et al. (Eds.): ICESD 2024, ESE, pp. 17–25, 2025.
https://doi.org/10.1007/978-3-031-88683-6_2

three criteria that must be met for a plant to be classified as a cadmium hyperaccumulator. Hyperaccumulator plants with cadmium contents in the aboveground parts should exceed 100 mg/kg. Meanwhile, the cadmium content in the aboveground parts should surpass that of in the underground parts and that of in the soil. In addition to demonstrating a robust capacity to accumulate heavy metals, hyperaccumulator plants should also exhibit a high degree of tolerance to these elements. Ideal hyperaccumulator plants should display the following characteristics: rapid growth, large biomass, robust resistance to pests and diseases, and the capacity to simultaneously accumulate two or more types of heavy metals [2].

The growth of hyperaccumulator plants is subject to variation as a consequence of the differing environmental factors, including soil type and climate, that are characteristic of different geographical regions. Therefore, it is meaningful to screen hyperaccumulator plants through outdoor soil pot experiments under specific regional and environmental conditions. The screening method is to sow the seeds of plants to be screened in seedling pots and transplant the seedlings into the polluted soil when they grow to a certain size, or collect some seedlings from specific families (such as Cruciferae, Poaceae, Asteraceae) and plant them in heavy metal-contaminated soil [2]. Once the plants have reached maturity, their physiological indicators and heavy metal content in various parts are measured to determine their tolerance and accumulation. When plants with good tolerance and accumulation are selected, a systematic heavy metal gradient experiment is conducted to determine if the plants have achieved hyperaccumulation capability for a particular heavy metal. This pot experiment method also holds significant practical implications. Another common screening method is to judge hyperaccumulator plants through field screening [3]. The soil in the vicinity of mining and smelting operations frequently exhibits elevated concentrations of heavy metals, and the plants cultivated in this soil demonstrate enhanced tolerance and accumulation of heavy metals. The plethora of plant species that populate the natural world represents a vast repository of potential for the remediation of heavy metal pollution. However, the plants that are ultimately selected for this purpose must demonstrate a robust capacity to withstand stress, adapt to diverse environmental conditions, and exhibit robust vitality. There are many problems with sampling and analysis in mining or polluted areas, such as difficulties in identifying plant species, uncertainty in collecting target plants, the risk of omitting too many accumulator plants, time-consuming and labor intensiveness, and difficulty for selected plants to grow under different climates and environments. Guo et al. employed soil from the Wudang area to identify accumulator plants capable of thriving in environments contaminated with elevated levels of lead and cadmium. Their study yielded Imperata cylindrica, Portulaca oleracea, and Phytolacca acinosa as promising candidates, which they subsequently validated through rigorous experimentation [4]. Therefore, conducting pot simulation tests using soil from specific regions for screening accumulator plants is a feasible method for screening hyperaccumulator.

2 Materials and Methods

2.1 Experimental Material

The test soil is collected from a cultivated land in Shuangliu District, Chengdu City and its physicochemical properties are shown in Table 1. This region has an altitude of 476 m and belongs to a subtropical humid monsoon climate. The climate of this region is characterized by mild conditions, with reduced sunlight and an extended period free of frost. The annual mean temperature is 16.3 °C, the annual mean precipitation is 855.8 mm, and the annual mean period of frost-free conditions is 289 days. The region is rich in plant growth, such as mulberry, clover, wild mugwort, humulus, fanzha, dogtail grass, barnyard grass, ferns etc. We collect surface soil, remove impurities, air-dry and sieve through a 5mm mesh for later use. The chemical purity of $CdCl_2.2.5H_2O$ has been established.

Table 1. Basic soil properties

PH	Organic matter (g/kg)	Electrical conductivity (us/cm)	Alkali-hydrolyzable nitrogen (mg/kg)
7.53	12.6	157.9	35.01
Quick-acting potassium (mg/kg)	Effective phosphorus (mg/kg)	Effective cadmium (mg/kg)	Total cadmium (mg/kg)
98.37	14.30	0.030	0.271

2.2 Experimental Design

We design pot experiments using plastic pots with an inner diameter of 20 cm and a height of 17 cm, equipped with trays. Each pot is filled with 1kg of soil sieved through a 5 mm mesh, and $CdCl_2.2.5H_2O$ is used as the cadmium additive reagent. In accordance with the findings of pertinent research and the outcomes of preliminary cultivation experiments [5–8], the cadmium addition levels for three gradients (A, B, and C) have been established at 0, 25, and 50 mg/kg, respectively. Urea and KH_2PO_4 are used as base fertilizers with dosages of 100 and 80 mg/kg, respectively. Mix the base fertilizer, cadmium and soil thoroughly. The small pots should be placed in the greenhouse for a period of one week to allow for maturation. Subsequently, after a period of one week, the plants should be watered to facilitate free growth. During this phase, it is essential to maintain the field capacity at approximately 70% by regular watering, as outlined in reference [9]. Harvest the plants three months later. After rinsing the plants thoroughly with tap water, rinse them several times with distilled water, dry them with filter paper. The plants should be subjected to a 105 °C withering process and subsequently dried at 75 °C until a constant weight is achieved. Thereafter, the plants should be weighed and pulverized into a powder. Measure the cadmium content in the plants. At the same time, collect soil samples, air-dry them naturally, grind them separately through 100 mesh and 60 mesh sieves, and measure the pH, electrical conductivity, available cadmium content, and total cadmium content in the soil.

2.3 Measurement Methods

Determination of soil basic Physicochemical Property

pH: measured using the potentiometric method (water-soil ratio is 2.5:1); Electrical conductivity: measured using the electrode method (water-soil ratio is 5:1); Alkaline nitrogen content: determined by semi-trace Kjeldahl method; Organic matter content: determined by the potassium dichromate volumetric (with external heating) method; Effective phosphorus: extracted using sodium bicarbonate leaching method; Quick-acting potassium: determined by ammonium acetate leaching method.

Determination of cadmium in plants

Weigh 0.2000 g of powdered plant sample, place it into 50 mL beaker and add 10 mL concentrated nitric acid and soak overnight. Transfer the sample from the beaker to a PTFE digestion vessel, and wash the beaker with 5 mL of nitric acid, ensuring that the wash solution is also transferred to the Teflon vessel without loss. The next step is to add 3 mL of hydrofluoric acid and 1 mL of perchloric acid to the digestion vessel, then cover and place it on a microcomputer-controlled temperature heating plate for digestion until almost dry. Once cooled, add 1ml of concentrated nitric acid to facilitate dissolution. Rinse with ultrapure water and filter into a 50ml volumetric flask to volume. Store in a refrigerator at 4 °C or analyze promptly using ICP-MS.

Determination of available cadmium

Refer to GB/T23739–2009 "Soil quality—analysisi of available lead and cadmium contents in soils".

Determination of total cadmium in soil

Weigh 0.2500 g of soil sample passed through a 100 mesh sieve into a 50 mL PTFE crucible, moisten with water, and add 5 mL of concentrated hydrochloric acid (excellent purity). Heat gently on a hot plate in a fume hood at low temperature (100 °C) to initiate sample decomposition. When it evaporates to about 2–3 mL, remove and let cool slightly, then add 5 mL of concentrated nitric acid (excellent purity), 10 mL of concentrated hydrofluoric acid (excellent purity), and 1 mL of concentrated perchloric acid (excellent purity). Cover the crucible and heat on the hot plate at medium temperature (180 °C) for approximately 1 h to drive off the acids. Continue dissolving until the solution is almost dry and viscous. Remove the crucible to cool and add 1 mL of concentrated nitric acid solution to dissolve the residue. Filter the solution into a 50 mL volumetric flask, make up to volume and shake well. Store in the refrigerator at 4 °C or immediately analyze by ICP-MS.

2.4 Data Processing and Analysis

We use Excel 2016, SPSS 22.0, and Origin 2017 for data analysis and processing. The main instruments and models are shown in Table 2.

Table 2. Test main instrumentation and equipment.

Instrument (Equipment) name	Model number	Instrument source
pH meter	pHS-25	Instruments and Electronics (Shanghai) Assosicates Co., Ltd.
Conductivity meter	DDS-11A	Shanghai Leichi Instrument Factory
Centrifuge	SC-3612	USTC Zonkia Scientific Instrument Co., Ltd.
Thermostatic oscillator	THZ-82A	Xiyuan Scientific Instrument Co., Ltd.
Adjustable electric-hot plate	ML-1.5–4	ZTE Vastitude Instrument Co., Ltd.
Digital display electronic thermostatic water bath	HH-2	Guohua Electrical Co., Ltd.
Electrothermal thermostatic blast drying oven	101A-2B	Shanghai Experimental Instrument Factory
Inductively Coupled Plasma Mass Spectrometer (ICP-MS)	M90	Bruker Corporation, USA
Atomic Spectral Absorptiometry	AA800	PerkinElmer Corporation, USA
Spectrophotometer	ALPHA-1506	Shanghai Jingsheng Scientific Instrument Co., Ltd.
Ultrapure water machine	WP-UP-11–10	Sichuan Voter Co., Ltd.
Electronic balance	BSA124S	Sartorius Scientific Instruments (Beijing) Co., Ltd.
Ultrasonic cleaner	JP-060	Skymen Technology Corporation Limited

3 Results and Analysis

3.1 Plant Species

Table 3 shows the screened plants. Clover, also known as Trifolium repens L., is a perennial herbaceous plant of the Fabaceae family, commonly used as forage and for landscaping. Fanzha, also known as Stellaria media (L.) Cyr., is an annual or biennial field weed in the caryophyllaceae family, widely distributed throughout the country. Humulus, also known as Humulus scandens (Lour.) Merr., belongs to the moraceae family and it is an annual or perennial creeping vine plant with vigorous growth and large biomass. Quinoa, also known as Chenopodium album L., is an annual herbaceous plant in the amaranthaceae family commonly found in fields and vegetable gardens. Dogwood, also known as Setaria viridis (L.) Beauv., is an annual herbaceous plant belonging to the poaceae family, commonly seen as a field weed. Barnyard grass, also known as Echinochloa crusgalli (L.) Beauv.), belongs to the gramineae family, is an annual herbaceous plant often found in rice fields.

Table 3. Basic characteristics both plant and soil in plant screening test.

Process number	Plant name	Dry weight (g)	Cadmium content in plants (mg/kg)	Available cadmium in soil (mg/kg)	Total cadmium in soil (mg/kg)	Enrichment factor (mg/kg)
A1	Chenopodium	3.11	0.40	0.19	0.53	0.75
A2	Chenopodium	2.91	0.37	0.17	0.58	0.64
A3	Dogwood	4.89	0.44	0.19	0.51	0.86
A4	Humulus	10.39	0.31	0.18	0.52	0.60
A5	Barnyard	4.18	0.43	0.16	0.48	0.90
B1	Clover	1.97	92.52	9.94	23.91	3.87
B2	Fanzha	0.94	99.68	10.14	24.41	4.08
B3	Clover	2.52	88.74	11.02	24.06	3.69
B4	Humulus	6.61	22.92	9.34	23.32	0.98
C1	Chenopodium	0.62	50.65	25.11	48.62	1.04
C2	Fanzha	0.33	103.03	24.01	47.21	2.18
C3	Chenopodium	0.76	41.79	24.36	46.82	0.89
C4	Humulus	2.98	56.93	26.08	47.84	1.19

3.2 Growth Characteristics of Different Plants

The basic characteristics both plant and soil in the plant screening test are shown in Table 3.

In treatment groups A4, B4, and C4, the dry weights of Humulus are 10.39, 6.61, and 2.98 g respectively, making it has the highest biomass among the six plants. However, the dry weight of Humulus shows a decreasing trend with increasing cadmium concentration. In A1 and A2 treatment, the dry weights of Chenopodium are 3.11 and 2.91 g, but the dry weights of Chenopodium are 0.62 and 0.76 g in C1 and C3 treatments, respectively, which shows the dry weights of Chenopodium significantly decreases with increasing in cadmium concentration. In B1 and B3 treatments, the dry weights of Clover are 1.97 and 2.52 g, respectively, with no discernible signs of toxicity. In B2 and C2 treatments, the dry weights of Fanzha are 0.94 and 0.33 g, respectively, representing the lowest values observed among the six plants. This limits the potential application of Fanzha in soil remediation. In A3 and A4 treatments, the dry weights of Dogwood and Barnyard are 4.89 and 4.18 g, respectively. However, these two species were not found in cadmium-amended soils.

In summary, the order of maximum biomass from highest to lowest are: Humulus > Dogwood > Barnyard > Chenopodium > Clover > Fanzha.

3.3 Heavy Metal Content Characteristics of Different Plants

From Table 3, it can be seen that the plant with the highest cadmium enrichment ability is Fanzha, with cadmium accumulation reaching 99.68 and 103.03 mg/kg in treatments B2 and C2, respectively, indicating a significant enrichment capacity for cadmium.In B1 and B3 treatments, the cadmium content in Clover reaches 92.52 and 88.74 mg/kg, respectively, showing a strong enrichment ability for cadmium, despite having the lowest biomass in its group. In A4, B4, and C4 treatments, the cadmium content in Humulus is 0.31, 22.92, 56.93 mg/kg, respectively. These results indicate that as the soil cadmium concentration increases, the cadmium content in Solanum nigrum also rises to a maximum at a soil cadmium addition of 50 mg/kg. In A1, A2, C1, and C3 treatments, the cadmium content in Chenopodium is 0.40, 0.37, 50.65, 41.79 mg/kg, respectively, which shows the cadmium content in Chenopodium increases with the cadmium concentration increasing. The cadmium content in dogwood and Barnyard is 0.44 and 0.43 mg/kg, respectively.

In summary, based on above statistics, the ranking of maximum cadmium content is as follows: Fanzha > Clover > Humulus > Chenopodium > Dogwood > Barnyard.

3.4 Enrichment Factor of Different Plants

From Table 3, it is known that the enrichment factor ranges from 0.27 to 4.08. In B2 and C2 treatments, the enrichment factor of Fanzha is 4.08 and 2.18, respectively, indicating a significant cadmium enrichment effect. However, as the concentration of cadmium increases, the enrichment factor decreases, indicating that concentrations exceeding a certain threshold reduce the capacity of Fanzha to enrich cadmium. In B1 and B3 treatments, the enrichment factor of Clover is 3.87 and 3.69, respectively, demonstrating its strong ability to enrich cadmium. The enrichment factor of Humulus is 0.60, 0.98, 1.19, respectively. These values demonstrate that as the cadmium concentration increases, the enrichment factor also increases, reaching its peak at a soil cadmium treatment concentration of 50 mg/kg. The enrichment factors of Chenopodium are 0.75, 0.64, 1.04, 0.89, respectively, also reaching its maximum cadmium concentration of 50 mg/kg.

In summary, based on above statistics, the ranking of maximum enrichment factor is as follows: Fanzha > Clover > Humulus > Chenopodium > Barnyard > Dogwood.

4 Discussion

The experimental results reveal that Humulus has a strong tolerance to cadmium-contaminated soil and is capable of growth in the three gradients of cadmium-contaminated soil. Wang Dayong et al. find that Humulus grows vigorously and has a strong enrichment capacity for cadmium in the investigation of heavy metal-polluted soil around the Fen River [10]. When the cadmium addition reaches 50 mg/kg, the cadmium content of Humulus reaches a maximum of 56.93 mg/kg, with an enrichment factor reaching 1.19. This indicates that Humulus has significant potential for the remediation of cadmium-contaminated soil. Furthermore, Wang Fengzhen et al. find that Humulus has good enrichment capability for cadmium as well [11]. In addition to its ability to enrich cadmium, Humulus also exhibits strong enrichment capabilities for copper and arsenic [12]. The cadmium content of Clover reaches a maximum of 92.52 mg/kg in the soil with a cadmium addition of 25 mg/kg, which is close to the critical value of hyperaccumulating plants and the highest enrichment factor reaches 3.87, demonstrating its significant enrichment capability for cadmium. This is similar to the research results of Yanjinshan et al. [6, 13]. As a kind of forage and landscaping plant, Clover demonstrates a high degree of environmental adaptability. Its absorption of cadmium is promoted by the strengthening effect of the biogenic chelating agent GLDA, rendering it valuable for further application and research.

Fanzha exhibits the highest maximum cadmium content and enrichment factor among these plants. Lin Lijin et al., in their study on Fanzha characteristics regarding cadmium enrichment, report that the maximum cadmium content in Fanzha reaches 135.79 mg/kg. However, they also observe a decrease in biomass as the cadmium content increases [9]. Zhao Gang et al. discove that Fanzha has good tolerance to various heavy metal-polluted soils and exhibits good enrichment capability for cadmium around landfill sites [14]. However, Fanzha has the lowest biomass in the soil with added cadmium, greatly limiting its application value. Chenopodium also exhibits a strong enrichment capability for cadmium but has relatively low biomass in the soil with added cadmium, which limits its application as well.

5 Summary

In light of the aforementioned experimental findings and analysis, the following conclusions can be drawn:

(1) Humulus, Clover, and Fanzha have strong tolerance to cadmium-contaminated soil and can grow well in the soil with high cadmium concentrations.
(2) At a soil cadmium content of 56.93 mg/kg, Humulus achieves a maximum enrichment factor of 1.19 for cadmium, indicating its significant potential for the remediation of cadmium-contaminated soil.
(3) Clover also demonstrates robust enrichment capabilities for cadmium, with a soil cadmium concentration reaching to 92.52 mg/kg, approaching the critical value of hyperaccumulating plants, and a maximum enrichment factor of 3.87.
(4) Fanzha exhibits strong enrichment capability for cadmium as well, with a soil cadmium content reaching up to 103.03 mg/kg and a maximum enrichment factor of 4.08. However, due to the low biomass of Fanzha, it is not easily cultivated and reproduced on a large scale.

Therefore, Clover and Humulus are chosen as the candidate plants for further research, considering both the plant biomass and their enrichment characteristics for cadmium.

Therefore, Clover and Humulus are chosen as the test plants for remediation cadmium-contaminated soil in future research, considering the plant biomass and cadmium enrichment characteristics.

References

1. Zhongmin, J.: Research on Phytoremediation Combination Technology Remedying the Soil Polluted with Leadand Cadmium. Northeast Forestry University (2013)
2. Shuhe, W., Qixing, Z., Xin, W., et al.: Screening of plants with heavy metal hyperaccumulation characteristics in weeds. Prog. Nat. Sci. **12**, 29–35 (2003)
3. Qixian, H.: Progress in screening of cadmium hyperaccumulating plants. Environ. Prot. Circular Econ. **33**(01), 46–49 (2013)
4. Longtao, G., Rong, F., Pinsha, Ou., et al.: Screening of Hyper-accumulation plant species in three types of habitat located inWudangdistrict. J. Guizhou Normal Univ. **33**(06), 23–28 (2017)
5. Jingwu, C.: Study on Response and Potential Phytoremediation of Bidens Pilosa L. Incadmium and Lead Stress. Southwest University (2013)
6. Yan, L., Ana, X.: Enrichment features of Trifolium pratense L.under cadmium stress. J. Henan Agric. Sci. **40**(01), 82–84 (2011)
7. Liping, S.: The Effects of Phyremediation and It's Combination Technology to Remediate Heavy Metal Cotaminated Soil by Ramie. Nanjing Agricultural University (2009)
8. Shouwen, Z.: Research on Phytoremediation of Soil Contaminated by Cadmium. Northwest Agriculture and Forestry University (2009)
9. Lijin, L., Bo, N., Mingan, L., et al.: Cadmium accumulation characteristics of winter farmland weed Stellaria media. Ecol. Environ. Sci. **23**(04), 673–678 (2014)
10. Dayong, W., Xingzhong, W., Zifang, Z., et al.: Origins of soil heavy metals and resistant plants around primary pollutant sources in the Linfen area of the Fenhe river. J. Shanxi Agric. Sci. **43**(10), 1290–1296 (2015)
11. Fengzhen, W., Xinjuan, S., Junhui, Xu., et al.: Absorption and accumulation of Heavy metals by aquatic plants in lakeside aone of Moshui Lake [J]. J. Wuhan Univ. Technol. **36**(11), 114–118 (2014)
12. Jujie, W., Yan, Z., Liuqing, C., et al.: Evaluation of heavy metal in soil and dominant plants of a copper mine waste-land in the center region of Hebei. Mineral Prot. Utilisation (1), 90–97 (2017)
13. Shan, J.: Remediation Potential of White Clover in Cadmium Pollution Soil. Northwest Agriculture and Forestry University (2013)
14. Gang, Z., huaiqing, W.: Hyperaccumulator plants and their heavy metal uptake in Qizishan Rubbish Landfill Disposal Site in Suzhou. J. Jiangsu For. Sci. Technol. **39**(6), 11–14 (2012)

Resource Management and Biodiversity Conservation

Collaborative Management of Water-Food-Energy Nexus Under Multiple Uncertainty

Yanxiao Zhou[1], Yongping Li[1,2(✉)], Guohe Huang[1,2], and Yanfeng Li[1]

[1] State Key Joint Laboratory of Environmental Simulation and Pollution Control, School of Environment, Beijing Normal University, Beijing, China
yongping.li@iseis.org

[2] Institute for Energy, Environment and Sustainable Communities, University of Regina, Regina, Sask, Canada

Abstract. Water shortage and food security have posed significant challenges to the sustainable management of water-food-energy (WFE) nexus. In this study, a fixed-mix joint-probabilistic constraints programming (FJCP) approach is developed for obtaining the sustainable WFE strategies. Then, a FJCP-WFE model is formulated with considering 36 scenarios. Results reveal that (i) system benefit and total water allocation would increase with the rising α as well as decreasing p levels, and decision makers can obtain the compromise among system violation risk, economic benefit and water consumption by varying p and α levels; (ii) the changing α level would influence water allocation structure, where the proportion of agricultural water would decrease by 4.9% from $\alpha = 0.47$ to $\alpha = 0$; (iii) the system benefit and total water allocation of FJCP would be lower than those of FCCP, with reductions of 0.001–0.04% and 1.92–5.32%, respectively. Results indicate that FJCP method is suitable for arid areas, which can find more practical alternatives for the WFE nexus system management, and promote regional sustainable development.

Keywords: fixed-mix · multiple uncertainty · water-food-energy nexus

1 Introduction

Due to the population growth and rapid socio-economic development, water shortage problems have been increasing prominent [1]. At present, over 1.8 billion populations are confronted with the risk of water scarcity, and 660 million people lack sufficient clean drinking water around the world [2]. Besides, water resources, food and energy systems are interrelated, and the shortage of water resources would endanger food safety as well as energy supply. On the contrary, crop cultivation and energy production would affect the amount and quality of water resources [3]. Therefore, the synergic management of water-food-energy (WFE) nexus system is necessary for achieving regional sustainable development. However, the WFE nexus system is afflicted with complex uncertainties [4], such as the food and energy demand would be random due to population fluctuation

G. Huang et al. (Eds.): ICESD 2024, ESE, pp. 29–38, 2025.
https://doi.org/10.1007/978-3-031-88683-6_3

as well as policy adjustment. Therefore, more robust optimizing approach need to be established for solving the above problems.

Joint-probabilistic constraints programming (JCP) is an operative approach in handling the random uncertainty of the WFE nexus system. For example, Zhang et al. [5] proposed an inexact JCP programming method for planning water resources [5]. Lv et al. [6] applied JCP method to studying the effect of energy and water shortage risks on the energy-water nexus system [6]. However, JCP cannot trace and modify the decision after random events occur. Fixed-mix stochastic programming (FSP) method can be utilized to deal with this problem, and reduce decision-making risk.

Therefore, a fixed-mix joint-probabilistic constraints programming (FJCP) method is presented for sustainable management of water-food-energy nexus system. The innovations and contributions include: (i) FJCP can examine correlations between food and energy systems; (ii) a FJCP-WFE model is presented with considering unconventional water utilization for the collaborative management of WFE nexus; (iii) discovering the optimal management plan for the Kai-kong river basin to guarantee food and energy security.

2 Methodology

A general FJCP approach can be expressed as [7]:

$$\mathrm{Max}\, f = \sum_{t=1}^{T} (c_t x_t - \sum_{k=1}^{K} p_{tk} d_t y_{tk}) \tag{1}$$

subject to:

$$\Pr\{a_{rt} x_t \leq b_{rt}\} \leq 1 - p,\ \forall r, t \tag{2}$$

$$a_{it} x_t + a_{itk}{}' y_{tk} \leq \omega_{itk},\ \forall i, t, k \tag{3}$$

$$x_{jt}, y_{jtk} \geq 0,\ \forall j, t, k \tag{4}$$

where p_{tk} is the probability of random event (e.g., water resource availability) under scenario k and period t, with $p_{tk} > 0$ and $\sum_{k=1}^{K} p_{tk} = 1$; K_t is the scenario number in period t; ω_{itk} is uncertain parameter corresponding to p_{tk}; x_t and y_{tk} are decision variables; p is joint constraint-violation probability. Through putting random variables under a series of separate probabilistic constraints, the FJCP method can be equivalently linearized to [8]:

$$\mathrm{Max}\, f = \sum_{t=1}^{T} \left(c_t x_t - \sum_{k=1}^{K} p_{tk} d_t y_{tk} \right) \tag{5}$$

subject to:

$$\Pr\{a_{rt} x_t \leq b_{rt}\} \leq 1 - p_i,\ \forall r, t \tag{6}$$

$$a_{it}x_t + a_{itk}{'}y_{tk} \le \omega_{itk}, \ \forall i, t, k \quad (7)$$

$$x_{jt}, y_{jtk} \ge 0, \ \forall j, t, k \quad (8)$$

$$\sum_{i=1}^{I} p_i \le p \quad (9)$$

3 Case Study

3.1 Statement of Problem

The Kaidu-Kongque river basin (abbreviated Kai-kong river) is located in Bayingolin Mongol Autonomous Prefecture. Kai-kong river (82°57″E–90°39″E, 40°25′N′–43°21″N) originates on the middle of Tianshan Mountain and flow into Bosten lake [9], which is mainly composed of Kaidu river, Bosten lake and Kongque river, with an area of approximately 62×10^3 km^2. This area belongs to an arid region, with an annual precipitation of 273 mm and evaporation of 1157 mm [10]. In recent years, with population growth and agricultural expansion, the Kai-kong river basin is facing serious water shortage and food security problems. Therefore, collaborative management of the WFE nexus system is crucial for easing water shortages and guaranteeing food security in Kai-kong river basin.

3.2 FJCP-WFE Modeling Formulation

Based on the developed FJCP method, a FJCP-based water-food-energy nexus management (FJCP-WFE) model is formulated for Kai-kong river basin. The model involves five planning periods, five water sources, twelve water users, and six types of energy conversion. Besides, in order to deal with the uncertainty of WFE model, three inflow levels, four utilization rate of brackish water, nine constraint-violation levels (as shown in Table 1) are considered. The FJCP-WFE model is formulated as:

$$\begin{aligned}
Max f = & \underbrace{\sum_{i}^{I}\sum_{k}^{K}\sum_{t}^{T} PEW_{kt} \times TAEW_{ikt}}_{\text{Benefit of energy sector}} - \underbrace{\sum_{i}^{I}\sum_{k}^{K}\sum_{t}^{T}\sum_{h}^{H} p_h \times CEW_{kt} \times \begin{pmatrix} TAEW_{ikt} - EDRW_{ikht} \\ -ESW_{ikht} - EGW_{ikht} - EBW_{ikht} \end{pmatrix}}_{\text{Economic penalty of energy sector}} \\
& + \underbrace{\sum_{i}^{I}\sum_{j}^{J}\sum_{t}^{T} PW_{ijt} \times TAW_{ijt}}_{\text{Benefit of other sectors}} - \underbrace{\sum_{i}^{I}\sum_{j}^{J}\sum_{t}^{T}\sum_{h}^{H} p_h \times CW_{ijt} \times \begin{pmatrix} TAW_{ijt} - SW_{ijht} - GW_{ijht} \\ -IRW_{ijht} - DRW_{ijht} - BW_{ijht} \end{pmatrix}}_{\text{Economic penalty of other sectors}} \\
& \underbrace{- \sum_{i}^{I}\sum_{j}^{J}\sum_{t}^{T}\sum_{h}^{H} p_h \times CE_{jt} \times \left(SW_{ijht} + DRW_{ijht} + BW_{ijht} + GW_{ijht}\right)/\gamma_{jt}}_{\text{Cost of water transport}} \\
& \underbrace{- \sum_{i}^{I}\sum_{j=1}^{4}\sum_{t}^{T}\sum_{h}^{H} p_h \times QF_{ijt} \times \left(SW_{ijht} + GW_{ijht} + DRW_{ijht} + BW_{ijht}\right)/(1 + LF_{ijt})/WPC_{ijt}}_{\text{Cost of agricultural production (including fertilizer, electricity, diesel and fixed costs)}}
\end{aligned} \quad (10)$$

Table 1. The cumulative distributions and corresponding values of random variables in different planning periods.

(p, p_1, p_2)	Electricity demand, 10^9 kWh					Food demand, 10^9 kg				
	t = 1	t = 2	t = 3	t = 4	t = 5	t = 1	t = 2	t = 3	t = 4	t = 5
(0.05, 0.01, 0.04)	11.76	12.81	13.70	14.59	15.17	3.32	3.40	3.48	3.59	3.71
(0.05, 0.025, 0.025)	11.49	12.55	13.51	14.30	15.99	3.35	3.46	3.48	3.63	3.79
(0.05, 0.04, 0.01)	11.32	12.43	13.49	14.17	14.91	3.45	3.56	3.67	3.79	3.95
(0.10, 0.01, 0.09)	11.76	12.81	13.70	14.59	15.17	3.24	3.07	3.13	2.87	3.29
(0.10, 0.05, 0.05)	11.19	12.31	13.28	14.09	14.83	3.32	3.36	3.44	3.54	3.66
(0.10, 0.09, 0.01)	10.17	11.74	12.67	12.88	14.28	3.45	3.56	3.67	3.79	3.95
(0.15, 0.05, 0.1)	11.19	12.31	13.28	14.09	14.83	3.23	2.96	2.99	3.05	3.12
(0.15, 0.075, 0.075)	10.52	11.95	12.91	13.74	14.54	3.27	3.18	3.23	3.31	3.41
(0.15, 0.1, 0.05)	9.42	11.33	12.30	13.23	13.95	3.32	3.36	3.44	3.54	3.66

Main constraints are as follows:

(1) Availability constraint of water resource

$$\begin{aligned}
&\sum_{i}^{I}\sum_{j}^{J} SW_{ijht}/\gamma_{jt} + \sum_{i}^{I}\sum_{k}^{K} ESW_{ijht}/\gamma_{kt} \le TSW_{ht} + \left(TSW_{ht-1} - \sum_{i}^{I}\sum_{j}^{J} SW_{ijht-1} - \sum_{i}^{I}\sum_{k}^{K} ESW_{ijht-1}\right)\\
&\sum_{i}^{I}\sum_{j}^{J} GW_{ijht}/\gamma_{jt} + \sum_{i}^{I}\sum_{k}^{K} EGW_{ijht}/\gamma_{kt} \le TGW_{t} + \left(TGW_{t-1} - \sum_{i}^{I}\sum_{j}^{J} GW_{ijht-1} - \sum_{i}^{I}\sum_{k}^{K} EGW_{ijht-1}\right)\\
&\sum_{i}^{I}\sum_{j}^{J} DRW_{ijht} + \sum_{i}^{I}\sum_{k}^{K} EDRW_{ikht} \le \sum_{i}^{I}\sum_{j=11}\sum_{t}^{T} \alpha_{ijt} \times \left(SW_{ijht} + GW_{ijht} + DRW_{ijht} + IRW_{ijht} + BW_{ijht}\right)\\
&\sum_{i}^{I}\sum_{j=10} IRW_{ijht} \le \sum_{i}^{I}\sum_{j=10}\sum_{t}^{T} \beta_{ijt} \times \left(SW_{ijht} + GW_{ijht} + DRW_{ijht} + BW_{ijht}\right)
\end{aligned} \tag{11}$$

(2) Utilization constraint of unconventional water resource

$$\begin{aligned}
&\left(\sum_{i}^{I}\sum_{j}^{J}\left(DRW_{ijht} + IRW_{ijht}\right) + \sum_{i}^{I}\sum_{k}^{K} EDRW_{ikht}\right) \Big/ \left(\sum_{i}^{I}\sum_{j=10}^{11}\left(SW_{ijht} + GW_{ijht}\right)\right) \ge PWR_{t}\\
&\left(\sum_{i=1}^{I}\sum_{j=1}^{J} BW_{ijht} + \sum_{i=1}^{I}\sum_{k=1}^{K} EBW_{ikht}\right) / TBA_{t} \ge PWB_{t}
\end{aligned} \tag{12}$$

(3) Food and energy demand constraint

$$\Pr\left[\begin{array}{l}
(ESW_{ikht} + EGW_{ikht} + EDRW_{ikht} + EBW_{ikht}) \times (\sigma_{ikt} + \varphi_{ikt} + \eta_{ikt}) \ge FED_{ikt}\\
\sum_{i}^{I}\sum_{j=1}^{4} YA_{ijt} \times \left(SW_{ijht} + GW_{ijht} + DRW_{ijht} + BW_{ijht}\right)/(1 + LF_{ijt})/WPC_{ijt} \ge \sum_{i}^{I} TFD_{it}
\end{array}\right] \ge 1 - p \tag{13}$$

(4) Electricity consumption constraint for agricultural sector

$$\begin{aligned}&\sum_{i}^{I}\sum_{j=1}^{4}\left(SW_{ijht}+GW_{ijht}+DRW_{ijht}+BW_{ijht}\right)/(1+LF_{ijt})/WPC_{ijt}\times\xi_{it}\times CELE_{t}\\&+\sum_{i}^{I}\sum_{j=1}^{4}\left(SW_{ijht}+GW_{ijht}+DRW_{ijht}+BW_{ijht}\right)/(1+LF_{ijt})/WPC_{ijt}\times YA_{ijt}\times EAGP_{jt}\leq TAED_{t}^{\max}\end{aligned} \quad (14)$$

(5) Diesel consumption constraint for agricultural sector

$$\sum_{i}^{I}\sum_{j=1}^{4}\left(SW_{ijht}+GW_{ijht}+DRW_{ijht}+BW_{ijht}\right)/(1+LF_{ijt})/WPC_{ijt}\times KLD_{jt}\leq TCD_{t}^{\max} \quad (15)$$

(6) Maximum power generation constraint

$$\begin{aligned}&(ESW_{ikht}+EGW_{ikht}+EDRW_{ikht}+EBW_{ikht})\times(\sigma_{ikt}+\varphi_{ikt}+\eta_{ikt})\leq PGC_{ikt}^{\max}\\&(ESW_{ikht}+EGW_{ikht}+EDRW_{ikht}+EBW_{ikht})\times\theta_{ikt}\leq PGC_{ikt}^{\max}\end{aligned} \quad (16)$$

(7) Water consumption constraint of energy sector

$$\sum_{k=1}^{K}(ESW_{ikht}+EGW_{ikht}+EDRW_{ikht}+EBW_{ikht})\leq TWE_{it} \quad (17)$$

4 Result and Discussion

Figure 1 presents system benefits and water allocation under 36 scenarios, the system benefits range from ¥2.44 × 10^{12} to ¥90.96 × 10^{12}. The highest system benefit corresponds to scenario 30 (S30) when $p = 0.05, p_1 = 0.04, p_2 = 0.01$ and $\alpha = 0.47$; the lowest system benefit occurs under S7 ($p = 0.15$, $p_1 = 0.05$, $p_2 = 0.10$, and $\alpha = 0$). In detail, system benefit would ascend along with decreasing p and increasing α. By different p levels, decision makers can attain the compromise between system violation risk and economic benefit. The highest water allocation would be 200.96 × 10^9 m^3 under S30 ($\alpha = 0.47$), and the lowest water allocation would be 178.71 × 10^9 m^3 under S2 ($\alpha = 0$), with a variation range of 0–12.46%. Result shows that water allocation would increase with the rising α level.

Figure 2 shows water allocation patterns for all users under multiple scenarios. Results present that the changing α levels would influence the water allocation structure. In detail, with the rising α level, the proportion of agricultural irrigation water would decrease from 51.3 to 46.4% when (p, p_1, p_2) = (0.05, 0.01, 0.04), and the share of resident would increase from 2.6 to 14.9%. This is mainly because the economic benefit of agricultural irrigation is lower than the other water users. Therefore, water resources would priority transfer to the users with low water consumption or high income with the increase of available water resources (i.e., rising α level). For example, with the increased α level (from S1 to S4), the total water allocation increased by 1.81 × 10^9 m^3, and the agricultural water increased by only 0.48 × 10^9 m^3. The change of agricultural water was only 26.37% of the total water allocation, which lead to a decrease in the proportion of agricultural water allocation.

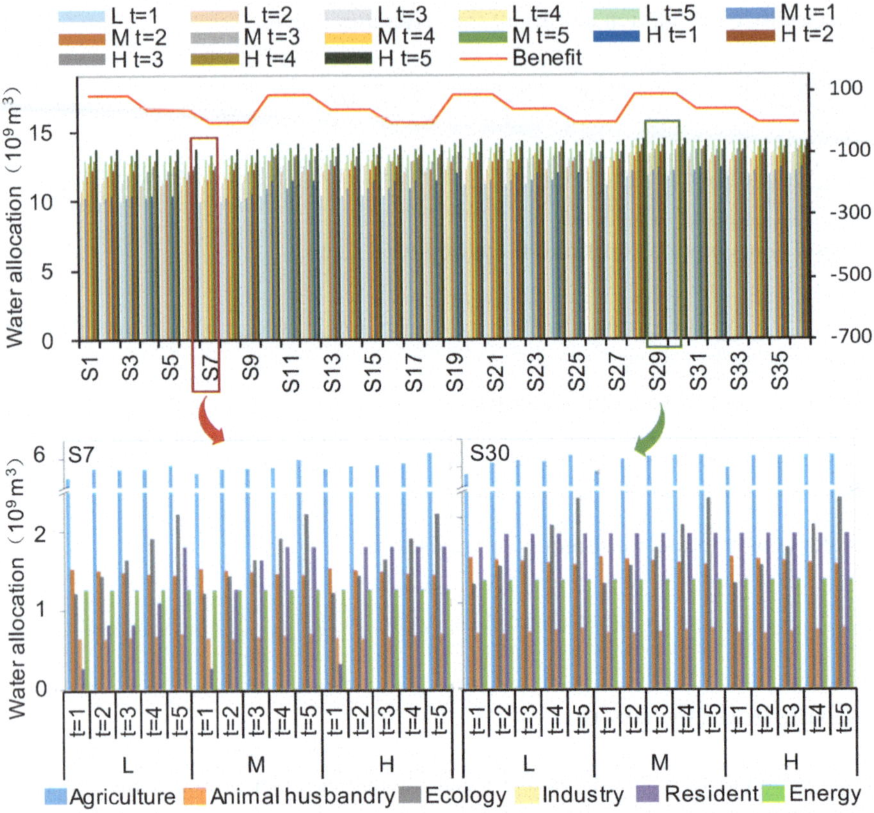

Fig. 1. Water allocation patterns and system benefits under all decision preferences.

Figure 3 presents the agricultural planting structure under different scenarios. Results indicate that the average annual planting area for Kai-kong river basin would be 148.53 $\times$ 10^3 ha to 178.63 $\times$ 10^3 ha, with the change of 20.27%. Besides, vegetable accounts for the largest proportion (i.e., 39.32–50.09%), followed by cotton. In order to further demonstrate the applicability of FJCP model for WFE nexus management, a traditional fixed-mix chance-constrained programming (FCCP) model was introduced for comparison with FJCP model. Table 2 shows the results of FCCP and FJCP model, indicating the system benefit and water allocation of FJCP would be lower than FCCP. Compared with FCCP, the system benefits from FJCP model would decrease by 0.001–0.04% under all scenarios; the total water distribution would reduce by 1.92–5.32%. This is mainly because the random variables of FCCP model are higher than those in FJCP, resulting in a significantly broader decision space. Therefore, FJCP is suitable for WFE management where random variables exhibit complex interactive relationships.

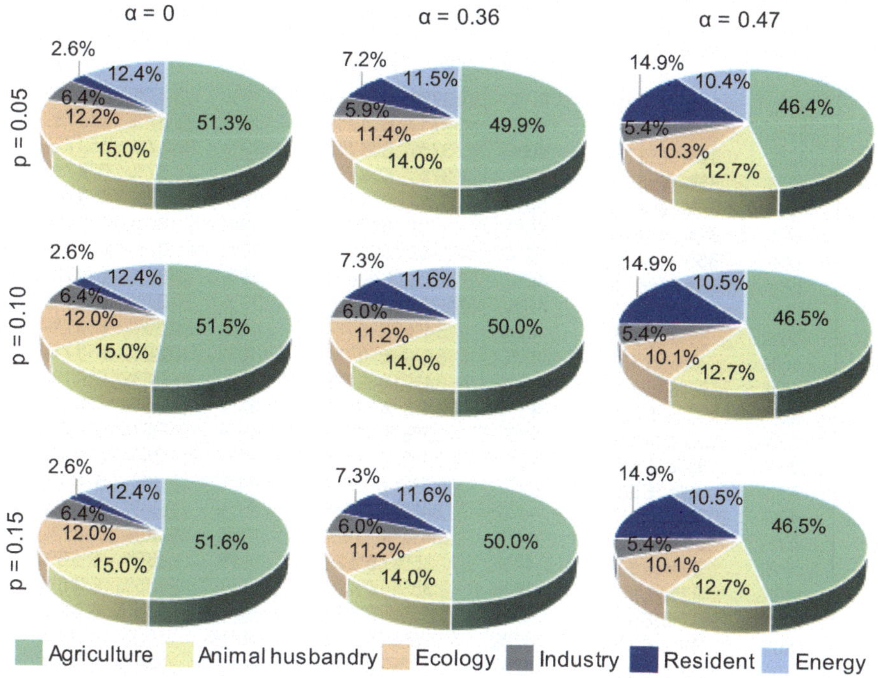

Fig. 2. The proportion of water allocation for different users.

5 Conclusions

A fixed-mix joint-probabilistic constraints programming (FJCP) method has been developed for water-food-energy (WFE) nexus management. FJCP can reflect the complex relationship between uncertain parameters and economic penalties. Then, the FJCP-WFE model has been applied to Kai-kong river basin, and 36 scenarios have been explored for handling random uncertainties in multiple subsystems. The results of FJCP-WFE model can help decision makers balance risk avoidance and system benefits, providing technical support for the sustainable development of water-food-energy nexus.

Some major findings are: (i) increasing p and falling α levels would lead to a decrease in system benefit and total water allocation; by different p and α levels, managers can achieve a compromise among water consumption, economic benefit and system violation risk; (ii) the share of water allocated for each department would be significantly affected by α level. When α from 0.47 to 0, the proportion of agricultural water would decrease from 51.3 to 46.4%, and that for resident would increase by 12.3%; (iii) the total water allocation and system benefit of FJCP would be lower than those of FCCP, for instance, the system benefit would reduce by 0.001–0.04%, and water allocation would reduce by 1.92–5.32%. Results present that lower water allocation obtained from FJCP could alleviate water contradiction between different sectors, providing support for the sustainable development of WFE nexus.

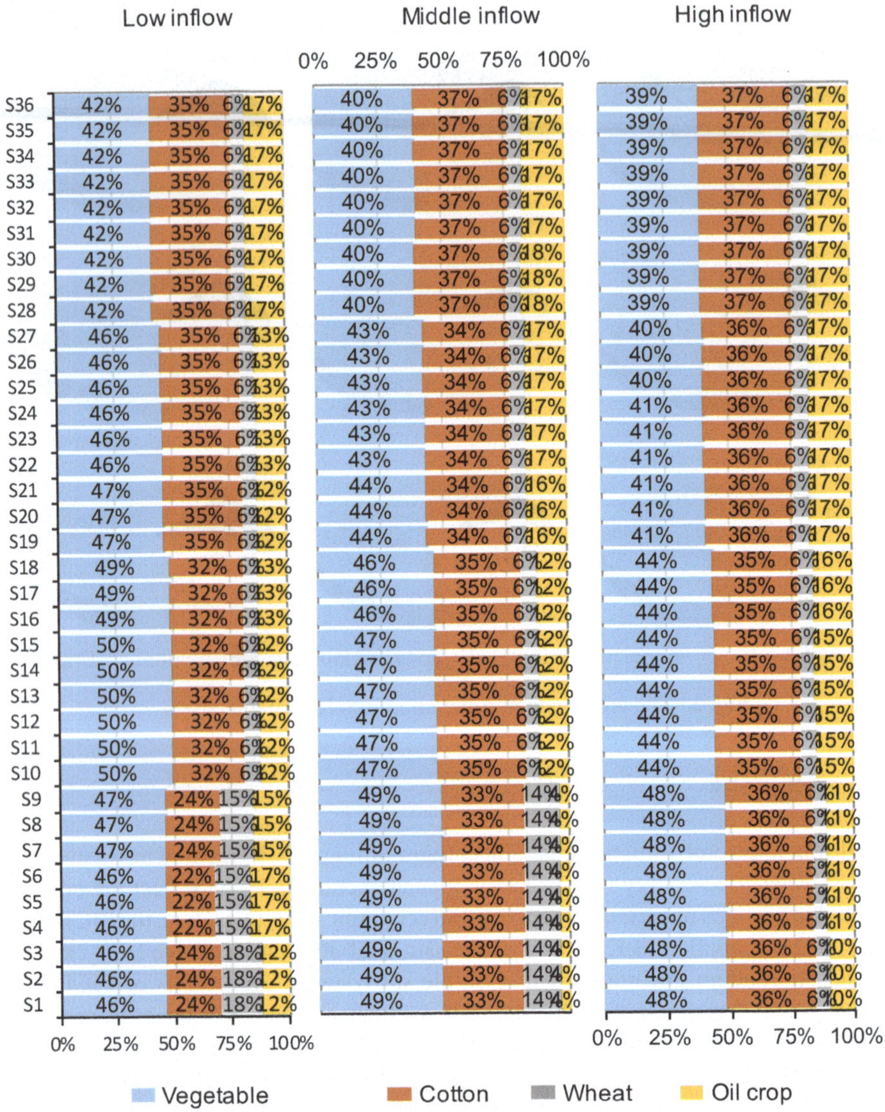

Fig. 3. The proportion of cultivated area for different crops under all planning schemes.

Table 2. Comparison of system benefit and water allocation from FCCP and FJCP.

p	System benefit (¥ 10^{12})		Water allocation (10^9 m^3)		p	System benefit (¥ 10^{12})		Water allocation (10^9 m^3)	
	FCCP	FJCP	FCCP	FJCP		FCCP	FJCP	FCCP	FJSCP
$\alpha = 0.47$					$\alpha = 0.36$				
0.05	90.96	90.96	67.26	65.81	0.05	90.95	90.95	62.99	61.21
		90.96		65.81			90.95		61.21
		90.96		65.81			90.95		61.21
0.1	41.08	41.08	67.13	65.78	0.1	41.07	41.07	62.92	61.16
		41.08		65.78			41.07		61.16
		41.08		65.78			41.07		61.16
0.15	2.50	2.50	67.02	65.73	0.15	2.49	2.49	62.86	61.12
		2.50		65.73			2.49		61.12
		2.50		65.73			2.49		61.12
$\alpha = 0.41$					$\alpha = 0$				
0.05	90.96	90.96	65.02	63.35	0.05	90.90	90.90	59.99	56.80
		90.96		63.35			90.90		56.80
		90.96		63.35			90.90		56.80
0.1	41.08	41.08	64.98	63.31	0.1	41.03	41.02	59.98	57.18
		41.08		63.31			41.02		57.18
		41.08		63.31			41.03		57.18
0.15	2.49	2.49	64.93	63.26	0.15	2.44	2.44	60.00	57.31
		2.49		63.26			2.44		57.31
		2.49		63.26			2.44		57.31

Acknowledgements. This research was supported by the National Key R&D Program of China (2023YFC3206503), Innovative Research Group of the National Natural Science Foundation of China (52221003), and the National Natural Science Foundation of China (52279003).

References

1. Tian, H., Ren, H., Li, X., Zhang, X., Xu, X., Wang, S.: Specificality, quality variation, assessment and treatment of estuarine water in the Pearl River Delta, South China. J. Environ. Inf. **42**(1), 13–24 (2023)
2. UNICEF/WHO (United Nations Children's Fund/World Health Organization).: Progress on sanitation and drinking water: 2015 update and MDG assessment (UNICEF/WHO)) (2015). https://www.who.int/publications/i/item/9789241509145
3. Peña-Torres, D., Boix, M., Montastruc, L.: Multi-objective optimization and demand variation analysis on a water energy food nexus system. Comput. Chem. Eng. **180**, 108473 (2024)

4. Nematian, J.: A two-stage stochastic fuzzy mixed-integer linear programming approach for water resource allocation under uncertainty in Ajabshir Qaleh Chay Dam. J. Environ. Inf. **41**(1), 52–66 (2023)
5. Zhang, C., Guo, S., Zhang, F., Engel, B.A., Guo, P.: Towards sustainable water resources planning and pollution control: inexact joint-probabilistic double-sided stochastic chance-constrained programming model. Sci. Total. Environ. **657**, 73–86 (2019)
6. Lv, J., et al.: Synergetic management of energy-water nexus system under uncertainty: an interval bi-level joint-probabilistic programming method. J. Clean. Prod. **292**, 125942 (2021)
7. Fleten, S.: The performance of stochastic dynamic and fixed mix portfolio models. Eur. J. Oper. Res. **140**, 37–49 (2002)
8. Gong, J.W., Li, Y.P., Lv, J., Huang, G.H., Suo, C., Gao, P.P.: Development of an integrated bi-level model for China's multi-regional energy system planning under uncertainty. Appl. Energy **308**, 118299 (2022)
9. Tarasov, P.E., et al.: An 8500-year palynological record of vegetation, climate change and human activity in the Bosten Lake region of Northwest China. Palaeogeogr. Palaeoclimatol. Palaeoecol. **516**, 166–178 (2019)
10. Mamat, A., Wang, J., Ma, Y.: Impacts of land-use change on ecosystem service value of mountain-oasis-desert ecosystem: a case study of Kaidu-kongque river basin, northwest China. Sustainability **13**(1), 140 (2021)

Biodiversity Net Gain Projects Under Environmental Change

Denis Vasiliev(✉)

Turiba University, Riga, Latvia
Deniss.Vasiljevs@turiba.lv

Abstract. Achieving biodiversity conservation targets in the context of exponential growth of anthropogenic land use expansion is highly problematic. The problem is globally relevant; however, it is particularly acute in the Global South where biodiversity is especially high, and the pace of development is unprecedented. Biodiversity Net Gain strategy is among the most promising approaches for addressing this issue, as it implies development to benefit rather than harm natural environment. This requires application of a holistic multi-scale approach that considers impacts of development on ecosystem networks. The networks also play a critical role in biodiversity adaptation to climate change. However, planning for biodiversity gain under changing climate can be highly problematic. Species distributions are changing, novel ecosystems are emerging, species responses to climate change differ and are often hard to predict. All this makes the already complex problem of achieving Biodiversity Net Gain objectives even more challenging. Therefore, the article explores ways of facilitating achievement of the objectives in the changing environment.

Keywords: sustainable development · Nature-based Solutions · landscape ecology · climate change

1 Introduction

Biodiversity decline has reached unprecedented levels with almost 70% of average relative species abundance loss across the Globe [1]. The decline is particularly sharp in tropical ecosystems, where in turn most of the world's biodiversity is concentrated [1]. At the same time, global infrastructure development projects are expanding globally, with particularly high intensity of the development expected in the Global South in upcoming decades [2]. This situation puts an enormous pressure on biodiversity as it is well established that infrastructure development drives species extinctions. In this context the approach to conservation has to focus on reconciling biodiversity protection and development objectives. A classic approach to conservation in the context of development projects is based on the Mitigation Hierarchy, which implies minimizing negative effects of the projects on the natural environment with subsequent restoration and recovery of ecosystems wherever possible [3]. The approach was long based on the idea of No Net Loss of Biodiversity; however, more recently a more promising approach aiming at net gain of biodiversity started gaining traction [4]. Biodiversity Net

G. Huang et al. (Eds.): ICESD 2024, ESE, pp. 39–48, 2025.
https://doi.org/10.1007/978-3-031-88683-6_4

Gain implies improving the state of biodiversity in the context of development projects [5]. Under this approach, ecosystem restoration and offsetting strategies are designed in such a way that intends to protect and restore more biodiversity than was lost as a result of development projects. In other words, this strategy involves both on-site and off-site conservation interventions. This requires holistic multi-scale view of the landscape and consideration of potential impacts of the environmental change.

The Biodiversity Net Gain strategy relies on a variety of funding sources depending on the local legal context. Biodiversity credit markets and carbon markets may be used to regulate funding of the projects. Often the projects involve a mix of public and private investment. The investors tend to expect some sort of guarantee of their investment outcomes. The fact that the projects are implemented across ecosystems at multiple spatial scales in the context of ecosystem complexity makes the outcomes highly uncertain. In addition to that, uncertainty increases due to the fact that the projects are usually implemented over long time periods in the changing environmental conditions. Both climate change and rapid land-use change can potentially undermine biodiversity outcomes. The only way to address change is to integrate it into the projects. However, the process of change management may be highly complex, and this may be the reason for its insufficient development. As a result, biodiversity goals may not be achieved. This may also make securing funding for the projects problematic as uncertain outcomes and lack of change management as a part of the planning process may affect investment decisions. Thus, there is a clear need for a robust approach to change management integrated into Biodiversity Net Gain strategies. In order to develop such approach, it is important to understand how both ongoing land use change and climate change may be addressed in the context of development projects aiming at biodiversity gain. Thus, the objective of this paper is to synthesize relevant knowledge and propose approaches for change management in the context of Biodiversity Net Gain strategies.

2 Land Use Change

Expansion of anthropogenic land uses to a large extent drives biodiversity loss. Recent findings suggest that more than 80% of species are endangered by land use change [6]. Land use change drive species extinctions through habitat loss and isolation. Development projects themselves actually tend to involve land conversion. In the context of Biodiversity Net Gain strategies, the habitat loss and fragmentation resulting from development is compensated either through restoration of sites disturbed during development of through ecosystem protection, restoration or creation elsewhere. In the long term, however, these sites may be affected by land use change and as a rest can lose capability to maintain biodiversity. Adjacent landscapes may be converted to anthropogenic land uses that are not permeable for species inhabiting the ecosystems that should deliver biodiversity value in the context of Biodiversity Net Gain strategies. Resulting disruption in landscape connectivity can undermine biodiversity due to limited gene flow and restriction of species ability to shift their range in a response to changing climatic conditions. Ecosystem isolation can also undermine its capacity to recover after major disturbances, making species assemblages within the ecosystem become less resilient, thereby potentially undermining their long-term persistence.

In order to ensure capacity of the restored, protected or developed ecosystems to maintain biodiversity it is essential to consider potential land use change scenarios during siting and planning stages of the conservation interventions. It is not a trivial task, however, modern technologies allow developing multiple scenarios considering a broad range of social and ecological factors simultaneously. Buffering the areas that are expected to deliver biodiversity benefits from negative impacts of land use change requires a holistic integrated approach to land use planning. The approach should not only consider sites dedicated to biodiversity conservation under Biodiversity Net Gain strategies but also broader landscape context. It is important to ensure that these sites are integrated into green networks that provide landscape connectivity at landscape and regional scales. In addition to that, it is essential to closely work with stakeholders that may have conflicting interests considering land use. Reconciling the conflicts during the siting and planning stages can sufficiently reduce the risk of future land use changes that may affect biodiversity on the sites selected for conservation. Additionally, collaboration with local landowners and integrated planning can help enhancing landscape characteristics, such as connectivity and heterogeneity, that can help to ensure capability of the sites selected for conservation under Biodiversity Net Gain strategies to maintain biodiversity. For instance, joint efforts of multiple farmers can substantially increase capacity of the landscape and ecosystems embedded in it to maintain biodiversity [7]. Thus, intertied approach based on multi-stakeholder collaboration, scenario planning and adaptive management needs to under implementation of Biodiversity Net Gain strategies (Fig. 1).

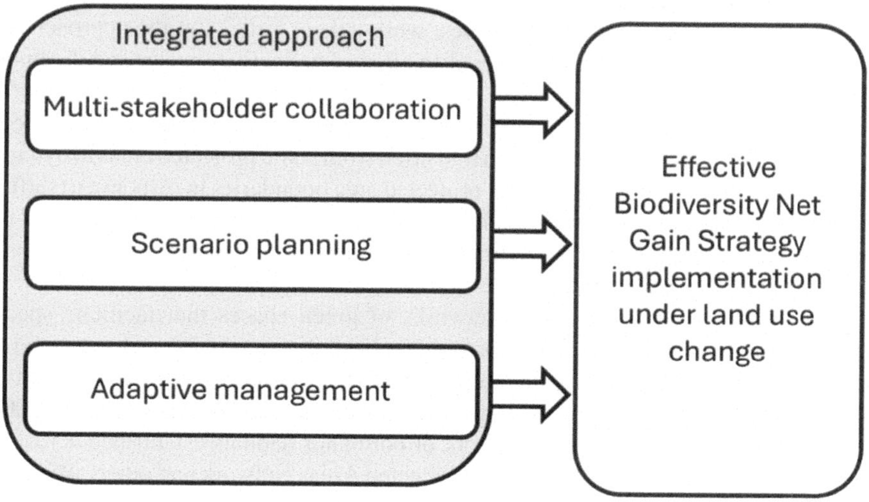

Fig. 1. Integrated approach to conservation planning for Biodiversity Net Gain under land use change.

3 Climate Change

Climate change is yet another source of uncertainty. Not only does it drive species extinctions but also it makes it harder to implement conservation measures. Species are shifting their ranges, new species assemblages emerge, interconnections between species are being disrupted and all this happens in the context of land use change, potentially impeding species responses to changing climate conditions by limiting their dispersal. All this means that establishing protected areas, restoring and developing ecosystems under Biodiversity Net Gain strategies based on current state of biodiversity and species distributions may be highly problematic as over the period of time since the sits are selected for protection, restoration or ecological engineering the state of biodiversity might dramatically change. In other words, habitats that are expected to support certain species over time may become located behind shifting ranges of the target species. At the same time, the ecosystems may become home for species that have never been present in the geographical location before. In addition to that plant species that lack mobility but at the same time form the basis of the protected, restored or developed ecosystem might not tolerate novel climatic conditions, ultimately leading to ecosystem collapse. These issues are hard to address, especially because climate change happens in the context of other environmental changes often acting in synergy. This situation leads to a lack of consensus in academic circles on how this situation should be addressed. Should the ecosystems be managed and designed for the future conditions or should local natural habitats be seen as a reference for the conservation sites? Indeed, the idea oof novel ecosystems is highly debated, with some scholars suggesting that we should embrace change and not refer to historical baselines, while others criticizing the approach as it is associated with certain risks such as proliferation of nonnative species and disruption of ecological networks [8].

The search for solutions is ongoing with many interesting and novel proposals constantly emerging. One such proposal is a transition from static protected areas to dynamic ones [9]. This approach implies shifting protected area boundaries in response to shifting species ranges and environmental conditions. The especially remarkable nature of this approach is that integrates well with offsetting strategy, which in turn can for the basis of Biodiversity Net Gain strategy.

Another approach is to develop networks of green spaces that facilitate species range shifts at a regional scale. The approach implies integration of the sites devoted for biodiversity conservation into local, national and transnational green networks. Some of the remarkable examples of the approach implementation include Natura 2000—protected area network, B-Lines—network of pollinator habitats established across the United Kingdom, Environment Canada's Protected Areas Network and others. While the approach is highly promising, the lack of cohesion within the network may ultimately undermine its effectiveness. For instance, some research has identified problems in some areas within European Natura 2000 network, especially in transboundary settings [10]. Nonetheless, the approach itself can be highly effective as can be seen from the projects implemented across England and Scotland, where habitat connectivity at a regional scale is often achieved through integration of green networks within the local landscape with regional networks. This approach, however, is based on integrated effort of multiple stakeholders. An interesting example of this is the Wendling Beck project in Norfolk where multiple farmers act in a coordinated manner to develop a rural habitat network, which would naturally integrate into the county's green network.

Whether or not it is possible to integrate habitats into climate proof networks question still remains, whether the site area can retain capacity to maintain biodiversity in the context of environmental change. There are many aspects to consider here. One is obviously species composition of the plants. Should the plants be selected considering future climate conditions and resilient to extreme weather evens, even if these plants are not native? An opinion exists that this approach may be less appropriate than selecting non-native and potentially invasive species. In addition to that it is also possible to utilize gene engineering techniques to use genetically modified species adaptive capacity of which could potentially be increased, making them more resilient for future climate change and to associated extreme weather events. Another important aspect to consider here is the area of the conservation site. A network of smaller habitats might have high level of species diversity, at the same time smaller habitats may be prone to stochastic events leading to species extinctions. Large habitat areas on the other hand may be more resilient to effects of climate change as well as buffer species from the negative effects of land use change. However, in larger habitats certain dominant species may suppress smaller populations of rare species which otherwise could potentially survive in smaller habitat patches. It is therefore essential to consider preservation of habitat size gradient within the landscape. Overall, a variety of strategies may be utilized for implementation of Biodiversity Net Gain under changing climate (Fig. 2).

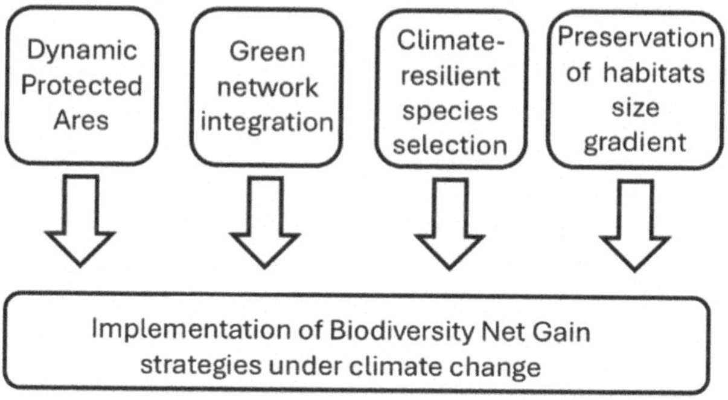

Fig. 2. Strategies that can be used for achievement of Biodiversity Net Gain under changing climate.

However, scarcity of empirical evidence proving effectiveness of each strategy challenges decision making. Further research is therefore essential to inform decisions made in the context of Biodiversity Net Gain strategy.

4 Synergistic Effects

A significant issue that merits particular attention is the synergy between negative effects of climate change and land use change. Ongoing landscape fragmentation associated with constant development leads to habitat isolation, which in turn restricts species adaptation to climate change through range shifts. Furthermore, not only species become restricted in their response to climate change through dispersal but also habitat isolation leads to decease in genetic diversity which in turn may reduce adaptive capacity, making species more vulnerable to negative effects of climate change. Implications of this problem for Biodiversity Net Gain are profound. Sites allocated for conservation under the strategy can become isolated over time due to expansion of anthropogenic land uses in adjacent territories. This could undermine their capacity to maintain biodiversity in the long term. Ultimately, Biodiversity Net Gain goals may not be achieved.

A range of strategies may be employed to address these problems. One approach could be to undertake thorough scenario planning and collaborative work with multiple stakeholders to try to secure their support and resolve possible future land use conflicts. This, however, may not be sufficient as in the changing world development pans may rapidly change. Thus, it is also essential to plan for possible future isolation. It is possible to buffer the ecosystem from negative future climate impacts by ensuring the presence of diverse micro-climatic conditions within its territory. There are multiple ways of achieving this, which include developing well connected tree and shrub habitats, conserving geo-diversity or integrating ecosystem management with wetland conservation. The choice of approach depends on landscape characteristics of the site. However, it is important to remember that planning for climate change resilience under possible land use change needs to take place and the situation should not be left to chance.

5 Effect Mitigation Means

Addressing the above described challenges should not discourage managers from pursuing Biodiversity Net Gain strategies. In the modern world numerous technological means are at the disposal of conservationists potentially allowing reducing the risks and uncertainty associated with the projects implemented in the context of change. It is well established that Artificial Intelligence and especially deep learning technologies have revolutionized modern science. Among the most powerful characteristics of these technologies are their capacity to help making predictions and simulate real-world systems, including highly complex natural ecosystems. In combination with Geographic Information System (GIS) technologies, these tools can be very powerful means or reducing uncertainty under environmental change. There are numerous applications for the tools that merit attention in the context of present discussion.

A most straightforward application of the artificial intelligence tools together with GIS is scenario development and planning. There are multiple options for application of the tools for this task. It is possible to use Long Short-Term Memory (LSTM) artificial recurrent neural networks in order to predict potential effects of climate change on biodiversity. In addition to that it is also possible to use convolutional LSTMs (convolutional networks underpin computer vision, thereby allowing working with images) with historic satellite imagery to develop potential land use change scenarios. For greater realism the analysis may be enhanced with individual-based modeling that can be based on deep reinforcement learning allowing simulating interactions of stakeholders as well as species responses to changing environmental conditions. Application of such sophisticated techniques may seem unaffordable in projects with limited budgets, especially those implemented in the Global South. However, software simplifying use of these technologies is already present and more user-friendly packages are likely to emerge in the near future. A remarkable example of recently developed software based on deep reinforcement learning is called Ecotwin [11]. Thiis tool was developed by Gothenburg University, Sweden. The tool allows loading a 3D terrain into the system and simulating biodiversity responses to different development interventions.

Another way to harness technological tooling for the sake of Biodiversity Net Gain Strategies is stakeholder management. As was mentioned earlier, mitigating potential negative impacts of land use change on the sites dedicated for biodiversity conservation can be achieved through close work and communications with key stakeholders. Here the first step is to identify who are the stakeholders to engage with. Identifying connections between people and potential influence that stakeholders have on each other can be facilitated through application of graph neural networks. The approach is used in marketing to analyze connections between stakeholders through analysis of the social media networks. This approach is very powerful and can greatly facilitate resolution of potential land use conflicts.

Less sophisticated technological tools can further facilitate work with stakeholders, especially with general public. Indeed, long term persistence of biodiversity within protected areas and restored or developed ecosystems to a large depends on public support. Land use change can potentially be driven by smallholder activities or through general public support for large development projects. Even establishment, restoration or development of sites dedicated for conservation may need to overcome resistance from local communities and secure their support. It is, therefore, critically important to engage with the general public effectively. The task is not trivial, due to the scale of communication activities and associated costs. Luckily, a large variety of tools have already been developed for modern marketing and can be readily used by conservation specialists. Numerous communication automation tools are available, including those allowing quicky generating content as well as distributing it across multiple communication channels. Furthermore, it is now possible to manage communications with the target audience with the help of Natural language processing tools based on deep neural networks that allow managing both written and oral communications (e.g., through application of chat-bots). The tools and their applications are summarized in Table 1.

6 Adaptive Management

The idea of adaptive management in conservation is well established practice [12]. It is often used to manage abrupt changes and tends to be based on the establishment of stakeholder networks. Considering application of the approach to Biodiversity Net Gain strategies in the context of environmental change, however, requires long term commitment and holistic view on social-ecological systems. This implies that the sites designated for biodiversity conservation have to be constantly monitored and managed according to changing environmental conditions. Several aspects need to be monitored, including biodiversity changes, changes in climatic conditions and trends inland use change on adjacent territories. This way it is likely to be possible to identify problems early, thereby potentially addressing them more effectively. Early warning system can

Table 1. The tools that can facilitate implementation of Biodiversity Net Gain projects.

Tool	Application
Geographic Information Systems (GIS)	Can be combined with artificial intelligence tools to model land-use changes, predict biodiversity outcomes, and identify vulnerable areas
Long Short-Term Memory (LSTM) Networks	Can help to predict long-term changes in species distributions and ecosystem functions under different climate scenarios
Convolutional LSTM Networks	Can be used with satellite imagery to predict land-use changes and biodiversity impacts, enhancing scenario planning
Individual-Based Modelling	Can be used to model species responses to changing environmental conditions, considering individual variability and interactions
Graph Neural Networks	Can facilitate stakeholder management by mapping social connections and influence, improving collaboration and reducing conflicts
Natural Language Processing	Can enhance public engagement through automated communication, content generation, and response management
Marketing Automation Tools	Can help engage the public and stakeholders, promoting support for conservation projects

All these tools are easy to use and many of them are also free

help take decisive actions that can buffer biodiversity from negative impacts of both climate change and land use change. For instance, radical actions such as species relocations are considered to be risky; however, this may be the only possible solution in the context of raid environmental change [13].

Vigilance is critical for successful implementation of adaptive management under Biodiversity Net Gain strategies. The question is, however, how to implement constant monitoring without excessive financial and labor investment. There are several things to consider here. First, engagement with local communities could be among the best ways to secure support for conservation action and help in constant biodiversity monitoring. Local people could be employed for protection, maintenance of the conservation sites and biodiversity monitoring. Furthermore, the broader community can be engaged in citizen science biodiversity surveys [14]. These approaches are highly beneficial as they build public support for conservation projects and at the same time tend to be highly cost effective. Secondly, the application of modern technologies can provide significant benefits. For instance, use of unmanned aerial vehicles can be successfully applied for real time biodiversity monitoring and can be particularly useful when the monitoring needs to be conducted over large areas. Indeed, modern unmanned aerial vehicles can be equipped with sensors that allow identifying not only large animals but even insects within the landscape, which can be particularly useful for evaluation of ecosystem health [15]. Importantly, the application of the vehicles is highly cost effective and can potentially be affordable even in the conservation projects with highly limited budgets. More technological tools can be used to facilitate adaptive management based on high vigilance. For instance, classic camera traps can be used together with computer vision technologies to automatically recognize species [16]. Another useful modern tool for highly effective biodiversity monitoring is bioacoustics [17]. Remarkably the technology is also relatively inexpensive but highly effective. Yet another modern approach to monitoring changes in species distributions and relative abundances includes Environmental DNA (eDNA) [17]. Along with biodiversity monitoring tools, application of freely available remotely sensed imagery combined with computer vision technologies can be utilized for timely change detection within the ecosystem as a whole. These changes may include ecosystem responses to changing climate conditions as well as incremental land use changes, which may initially be hard to detect. Overall, modern technologies are rapidly developing constantly offering new easy to use and affordable tools for conservation.

7 Conclusions

Implementation of Biodiversity Net Gain strategies is challenged by the ongoing environmental change. Climate change and land use change separately and in synergy can fundamentally undermine biodiversity outcomes of conservation actions implemented under the strategies. This fact should not discourage managers from implementation of the Biodiversity Net Gain. However, it is critically important not to ignore the fact that the projects are implemented in the context of change. Rather, strategies to mitigate potential impacts of environmental change need to be integrated into planning and implementation processes. This way biodiversity outcomes should be secured not only in short term but also in longer term. Location Biodiversity conservation sites should be selected considering potential impacts of the change. The planning process should involve integration of mitigation measures into the site territories. Furthermore, ongoing monitoring and adaptive management need to take place already after the projects are implemented. Thanks to modern technologies all this has become increasingly affordable to managers. Thus, opportunities to utilize technological tools should be considered. Further research on ways to mitigate impacts of environmental change on conservation projects implemented under the Biodiversity Net Gain strategies is urgently needed.

References

1. Almond, R.E., Grooten, M., Juffe Bignoli, D., Petersen, T.: Living Planet Report 2022–Building a Nature Positive Society. WWF, Gland, Switzerland (2022)
2. Meijer, J.R., et al.: Global patterns of current and future road infrastructure. Environ. Res. Lett. **13**, 064006 (2018)
3. Arlidge, W.N.S., et al.: A global mitigation hierarchy for nature conservation. **68**(5), 336–347 (2018)

4. Bull, J.W., Brownlie, S.: The transition from no net loss to a net gain of biodiversity is far from trivial. Oryx **51**(1), 53–59 (2017)
5. CIEEM, Biodiversity Net Gain: https://cieem.net/i-am/current-projects/biodiversity-net-gain/. Last accessed 31 May 2024
6. Hald-Mortensen, C.: The main drivers of biodiversity loss: a brief overview. J. Ecol. Nat. Resour. **7**(3), 000346 (2023)
7. Landis, D.A.: Designing agricultural landscapes for biodiversity-based ecosystem services. Basic Appl. Ecol. **18**, 1–12 (2017)
8. Miller, J.R., Bestelmeyer, B.T.: What's wrong with novel ecosystems, really? Restor. Ecol. **24**(5), 577–582 (2016)
9. Bull, J.W., et al.: Conservation when nothing stands still: moving targets and biodiversity offsets. Front. Ecol. Environ. **11**(4), 203–210 (2013)
10. Wood, S.L.R., et al.: Missing interactions: the current state of multispecies connectivity analysis. Front. I Ecol. Evol. **10**, 830822 (2022)
11. Ecotwin. https://www.ecotwin.se/. Last accessed 07 June 2024
12. Folke, C., et al.: Adaptive governance of social-ecological systems. Annu. Rev. Environ. Resour. **30**, 441–473 (2005)
13. Butt, N., et al.: Importance of species translocations under rapid climate change. Conserv. Biol. **35**(3), 775–783 (2020)
14. Conrad, C.C., Hilchey, K.G.: A review of citizen science and community-based environmental monitoring: issues and opportunities. Environ. Monit. Assess. **176**, 273–291 (2011)
15. Xavier, S.S., et al.: Remotely estimating beneficial arthropod populations: implications of a low-cost small unmanned aerial system. Remote Sens. **10**(9), 1485 (2018)
16. Schneider, S., et al.: Past, present and future approaches using computer vision for animal re-identification from camera trap data. Methods Ecol. Evol. **10**(40), 461–470 (2019)
17. Beng, K.C., Corlett, R.T.: Applications of environmental DNA (eDNA) in ecology and conservation: opportunities, challenges and prospects. Biodivers. Conserv. **29**, 2089–2121 (2020)

Generation of Monthly Flows in the Ica River Basin up to the La Achirana Hydrometric Station Using the GR2M Precipitation-Runoff Model

Angel Gabriel Ochante Sanchez(✉), Diego Fermin Jonislla Rojas, and Abel Carmona Arteaga

Universidad Privada del Norte, Lima, Peru
{N00261508,N00288531}@upn.pe, abel.carmona@upn.edu.pe

Abstract. Due to historical problems, such as river overflows in the Pacific slope basins, this study focuses on the Ica basin, which has recently experienced climatological disasters. The main objective is to generate monthly flows over a period of 32 years using the GR2M model, of which we will also explain its limitations, from the basin to the La Achirana hydrological station. Results showed an R2 of 0.85 and a Nash-Sutcliffe efficiency of 0.78, indicating a good fit between observed and simulated data. The GEE platform was used to obtain temperature (ERA5) and precipitation (RAIN4PE) data. The calibration and validation of the model was carried out using daily flow data (monthly averages) from the Water Observatory (ANA) of the La Achirana hydrological station between September 2013 and December 2015. This methodology aims to offer a comprehensive examination of the interaction between climatic variables and river flows, with important implications for the management and forecasting of extreme hydrological events.

Keywords: Basin · ERA5 · GEE · GR2M · La Achirana · RAIN4PE

1 Introduction

In Peru, the inhabitants of places near rivers have been affected by constant problems related to the water resources of their area. A particular and worrying case is the Ica River, which in recent years has presented strong flows and, consequently, climatological disasters such as overflows (See Fig. 1).

Although it would be expected that, after quite some time, as early as 2024 the authorities will face these problems. On the night of Wednesday, January 24, the Ica River overflowed after reaching its maximum capacity (See Fig. 2).

G. Huang et al. (Eds.): ICESD 2024, ESE, pp. 49–69, 2025.
https://doi.org/10.1007/978-3-031-88683-6_5

El Comercio

ICA / News

Overflow of the Ica River affected more than 100 hectares of crops

Injured farmers ask that the Regional Government provide them with heavy machinery to be able to reinforce the banks of the river. The governor of Ica supervised the affected area last night.

More than 100 hectares of crops were affected by the overflowing of the Ica River. (Photo: Gore Ica) **01/13/2021 12H11**

Fig. 1. News about the overflow of the Ica River that affected more than 100 hectares of crops [1].

This problem persists a month later. The district of Ocucaje, located in the province of Ica, experienced a slight increase in the flow of the Ica River, resulting in the overflow of the Braulio Channel [3].

What strategies or actions can be implemented or planned in order to mitigate this phenomenon?

One possible solution would be to erect reservoir infrastructure in the mountainous areas of certain river basins, with the aim of meeting the water needs of downstream communities, while regulating the extreme flows that could occur.

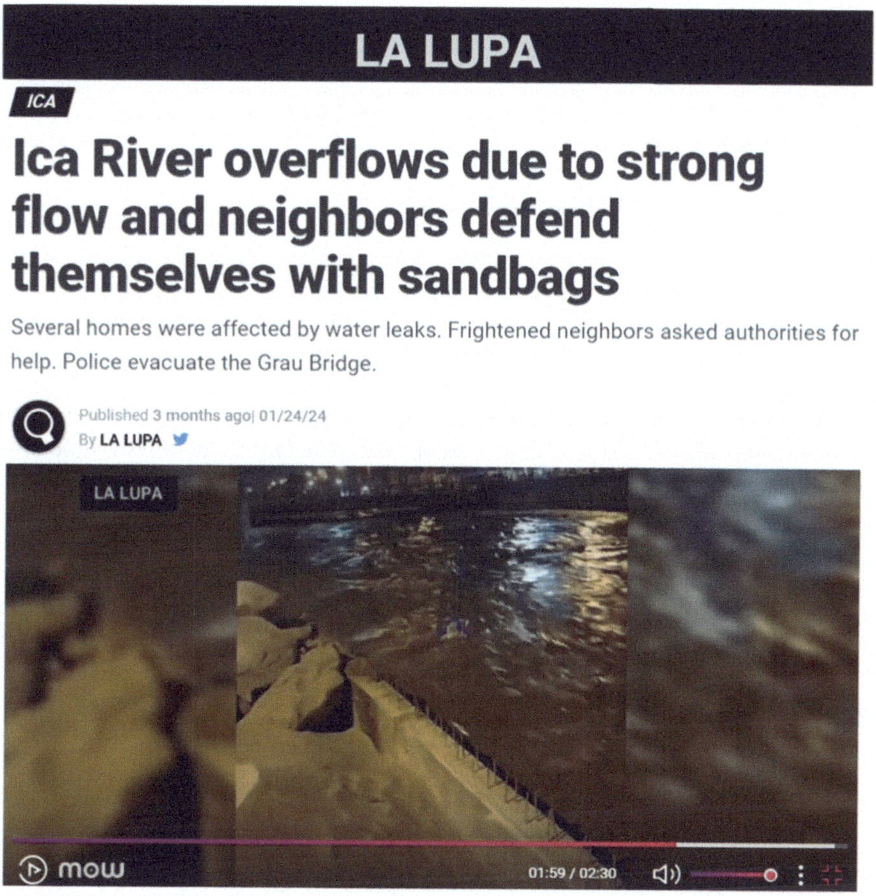
LA LUPA

ICA

Ica River overflows due to strong flow and neighbors defend themselves with sandbags

Several homes were affected by water leaks. Frightened neighbors asked authorities for help. Police evacuate the Grau Bridge.

Published 3 months ago| 01/24/24
By LA LUPA

Fig. 2. Local news reporting the overflow of the Ica River due to strong flow [2].

2 Research Objective

Given the context presented, and considering the limited availability of hydrometeorological data in the Ica River basin, this study proposes the generation of series of average monthly flows in this basin up to the La Achirana hydrometric station, covering the period between 1981 and 2013. This objective will be carried out through the application of the GR2M precipitation-runoff model [4]. For example, the GR2M conceptual model has been widely used in different hydroclimatic conditions around the world with satisfactory results and even better performance than other water balance models [5]. In addition, it is used to evaluate the effects of climate change on water resources. For this model, the use of gridded climate data of precipitation (RAIN4PE) and temperatures (ERA5-Land) is contemplated.

3 Theoretical Framework

3.1 Maximum Flow in Hydrological Models

The concept of maximum flow is essential to understand and manage river basins, especially in flood-prone regions where hydraulic works such as river defenses, dams, etc. could be carried out. This section reviews recent studies and methodologies related to peak flow estimation and its implications for hydrological modeling and water resources management.

Halil and Hafzullah (2020) demonstrate that the model's ability to simulate flow variations can be adapted to the Ica River basin. This adaptation can improve flood forecasting and water management by providing a reliable method for estimating monthly flow duration, thereby assisting in planning, construction implementation of hydraulic projects, and implementation of flood prevention programs [6].

The methodology proposed by Traverso et al. (2019) can be adapted to the Ica River basin to improve the accuracy of flow simulations. By incorporating diverse hydrological inputs, this approach can improve predictive accuracy, supporting better planning, water resource management, as well as more effective flood prevention measures and delivery of civil construction projects such as riverine defences [7].

3.2 Study Area Location

The Ica Basin region is bordered to the north by the Pisco Basin and to the east by the Cuenca Grande. Its territorial extension covers 7340.86 Km^2.

The hydrometric station is located at geographical coordinates of longitude −75.6833 and latitude −13.9667 [8].

3.3 Cartographic Data

The information was extracted from the national charts of the National Geographic Institute (IGN) on a scale of 1:100,000, which were used to delimit the area of the Ica Basin. The sheets of the National Chart used for the delimitation of the Ica River Basin are: 27-m, 28-l, 28-m, 28-n, 29-l, 29-m, 30-l, 30-m.

In the Ica Basin, a range of inclinations is identified that vary from its highest point, located at 131%, to its lowest point, which is at 20% (See Fig. 3).

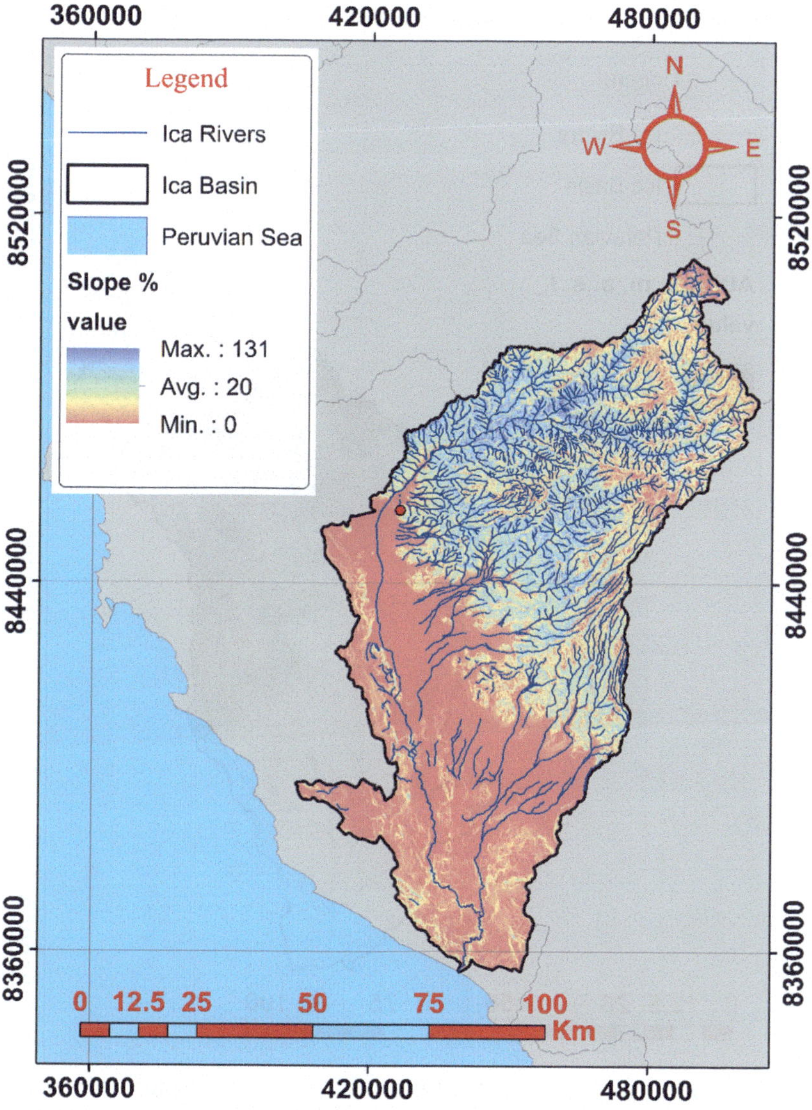

Fig. 3. Distribution of slopes in the Ica basin.

3.4 Morphological Data of the Basin

The Ica Basin covers an area of 7340.86 km^2. It can be seen in Fig. 2 that the distribution of elevations in the basin ranges from 4650 m.a.s.l. in the high parts and 0 m.a.s.l. in the lowest parts; In rugged parts the flow of water varies (See Fig. 4).

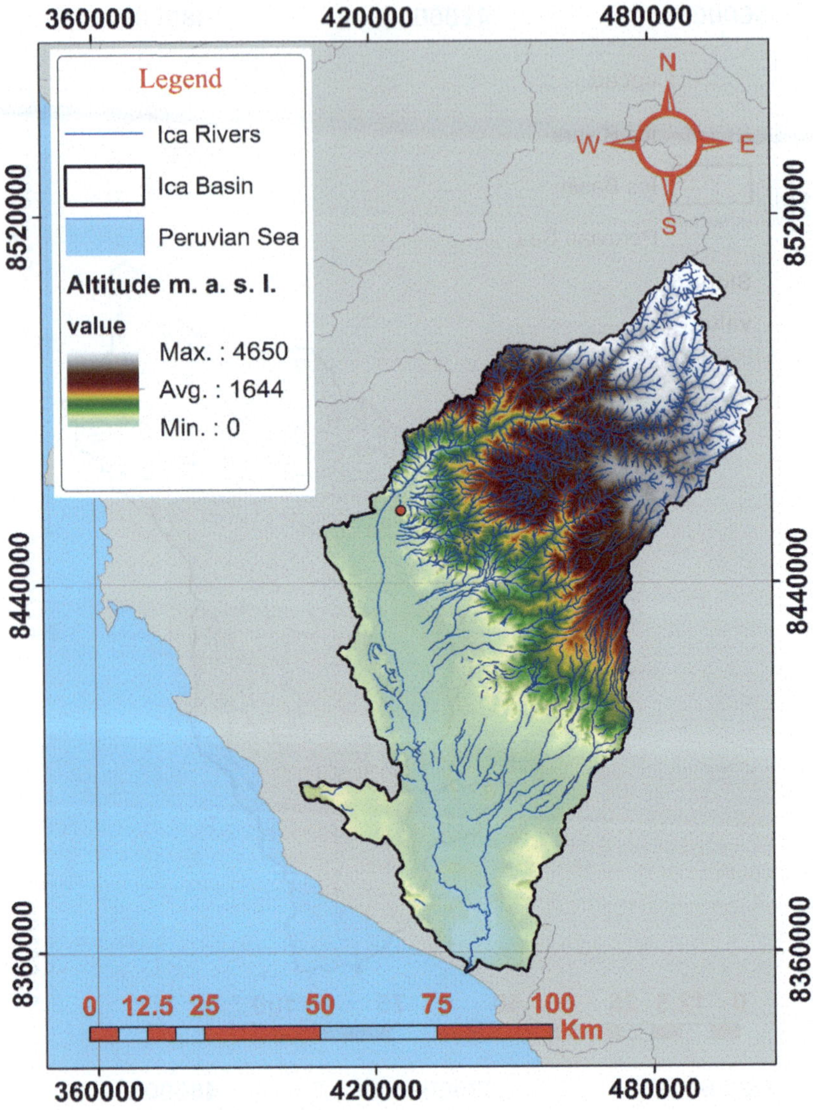

Fig. 4. Altitude distribution in the Ica Basin.

3.5 ArcGIS

It is a geographic information system (GIS) developed by the Environmental Systems Research Institute (ESRI), which allows collecting, organizing, managing, analyzing, sharing and distributing geographic information [9].

3.6 National Geographic Institute (IGN Acronym in Spanish)

The IGN is the governing body of national cartography; it has the mission of developing and updating the official basic cartography of Peru.

For the delimitation of the hydrographic basins, the cartographic information from the national charts 27-m, 28-l, 28-m, 28-n, 29-l, 29-m, 30-l, 30-m was used, which were They are located at coordinates WGS84, UTM—Zone 18 South.

3.7 Google Earth Engine (GEE)

It is a computing platform that allows users to run geospatial analyzes on Google infrastructure [10].

To obtain this geospatial analysis information, in this case, meteorological data, codes had to be created, which were provided by the Mg. Sc. Abel Carmona Arteaga [11].

Code for grilled product RAIN4PE. To obtain meteorological data on precipitation and average temperatures, the GEE platform was used, which is an innovative cloud-based geomatics application that allows viewing and analyzing multispectral images collected by satellites orbiting the planet Earth.

The code provided by the Mg is described below. Sc. Abel Carmona Arteaga:

```
// rain4pe: High resolution gridded precipitation dataset for Peruvian and Ecuadori-
an basins (1981-2015)
  // Global parameters
  var ROI = ee.FeatureCollection(projects/ee-
angelgabrielochantesanchez/assets/ICA_BASIN_LA_ACHIRANA_STATION');
  var start_date = "1981-01-01";
  var end_date = "2015-12-01";
  // Function to calculate the spatial average
  var ppreducer = function(rain4pe) {
  var params = {collection: ROI, reducer: ee.Reducer.mean(), scale: 5000};
  var image_value = rain4pe.reduceRegions(params).first().get('mean');
  var image_date = rain4pe.get('system:time_start');
  var ft = ee.Feature(null, {'system:time_start': image_date,
  'date': ee.Date(image_date).format('Y-M-d'), 'value': image_value});turn ft;};
  // Create a time series chart
  var rain4pe_daily_data = ee.ImageCollection("projects/sat-io/open-
datasets/rainpe/daily")
  var rain4pe_data = rain4pe_daily_data.filterDate(start_date,
end_date).map(ppreducer);
```

```
var graph = ui.Chart.feature.byFeature(
  rain4pe_daily_data.filterDate("1981-06-01", "1983-06-
01").map(ppreducer),'system:time_start','value');
print(graph.setChartType("ColumnChart").setOptions({vAxis: {title: 'PP
(mm/dia)'},hAxis: {title: 'fecha'}}));
// Export results
export.table.toDrive({collection: rain4pe_data,selectors: 'date, value',description:
"rain4pe-daily data", fileNamePrefix:"rain4pe_ts"})
// Show a specific day
var daily_rain = ee.Image("users/ryali93/rainpe/daily/1998_01_25")
var palette = ['000096','0064ff', '00b4ff', '33db80', '9beb4a','ffeb00', 'ffb300',
'ff6400', 'eb1e00', 'af0000'];
Map.addLayer(daily_rain, {min:0, max:80, palette: palette})
Map.centerObject(ROI);
Map.addLayer(ROI)
```

Code ERA5. The creator of the Google Earth Engine code for data extraction of the ERA5 project is Mg. Sc. Abel Carmona Arteaga [12].

```
// Define periods
var param = {
T1:'2015-12-01',
T0:'1981-01-01',
}
// Basin analyzed in Shapefile format
var Station = ee.FeatureCollection("projects/ee-
angelgabrielochantesanchez/assets/ICA_BASIN_LA _ACHIRANA_STATION")
//var Station = geometry
// Collect data and filter by total dates
var Precipitation = ee.ImageCollection('ECMWF/ERA5/DAILY');
//var Precipitation = ee.ImageCollection('UCSB-CHG/CHIRPS/DAILY')
// Wind data collections for all years, one value per hour
// This is where you can change the dates
var collection01 = ee.ImageCollection(Precipitation.filterDate(param.T0,
param.T1));
// Clip to specified region
var clipped01 = collection01.mean().clip(Station)
// graphics //
// Long-term time series
var TS1 = ui.Chart.image.seriesByRegion({
imageCollection: collection01,
.
reducer: ee.Reducer.mean(),
// Change data source //
// To use Copernicus average air temperature at 2m in Kelvin
band:'mean_2m_air_temperature',
scale: 5000,
```

```
    xProperty: 'system:time_start',
    seriesProperty: 'Subbasin'})
    .setOptions({
    title: Average precipitation in a basin P(meters) at daily rate',
    hAxis: {title: 'Time interval in days'},
    vAxis: {title: 'P (meters)'}}
    )
    ;
    print(TS1);
  Map.addLayer(Station,{palette: 'blue'}, 'basin')
```

Emphasize that precipitation is obtained in millimeters (mm) and temperature in Kelvin (K), which had to be converted to Celsius.

3.8 National Water Authority (ANA Acronym in Spanish)

It is the governing body and the highest technical-regulatory authority of the National Water Resources Management System of Peru [13].

3.9 RAIN4PE Grilled Product

RAIN4PE is a new gridded precipitation data set and product for Peru and Ecuador that uses the largest available in-situ observations, elevation data, and additional streamflow data to correct for precipitation underestimation.

RAIN4PE can be adopted as a reference precipitation product to evaluate other precipitation products in Peru and Ecuador.

RAIN4PE can be used for multiple hydrometeorological applications such as: water balance, spatio-temporal analysis of droughts and floods, etc. (See Fig. 5) [14].

3.10 Era5

It provides global coverage atmospheric information generated on an hourly and daily basis, resulting in updated and improved estimates of the state of the atmosphere [15].

3.11 Model GR2M

It is a French model that simulates flows on a monthly time scale. It has been used as a sequential simulation of soil humidity and flow data in conceptual precipitation-runoff models, obtaining very satisfactory results [16]. The GR2M model is a monthly LHM [4].

In Figs. 5 and 6, a diagram of the GR2M model algorithm can be seen. The hydrological model uses the Nash-Sutcliffe coefficient as an objective function in the calibration pase. (See Fig. 7 and the next 12 equations).

$$S_1 = \frac{S + X_1\varphi}{1 + \varphi\frac{S}{X_1}} \tag{1}$$

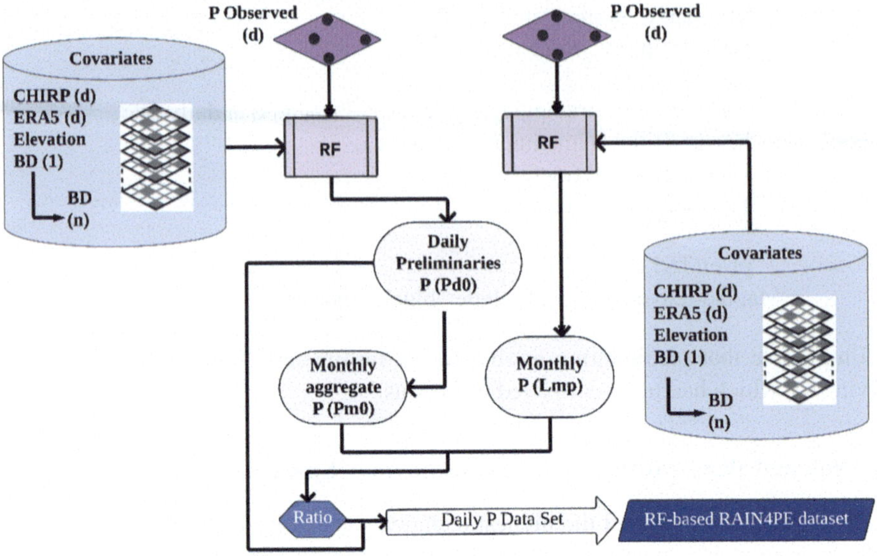

Fig. 5. Merging multi-source precipitation (P) data using random forest (RF) [14].

$$\varphi = \tanh\left(\frac{P}{X_1}\right) \tag{2}$$

$$P_1 = P + S - S_1\varphi \tag{3}$$

$$S_2 = \frac{S_1(1-\psi)}{1+\psi\left(1-\frac{S_1}{X_1}\right)} \tag{4}$$

$$\psi = \tanh\left(\frac{E}{X_1}\right) \tag{5}$$

$$S = \frac{S_2}{\sqrt[3]{1+\left(1-\frac{S_2}{X_1}\right)^3}} \tag{6}$$

$$P_1 = S_2 - S \tag{7}$$

$$P_3 = P_1 + P_2 \tag{8}$$

$$R_1 = R + P_3 \tag{9}$$

$$R_2 = X_5 - R_1 \tag{10}$$

$$Q = \frac{R_2^2}{R_2 + 60} \tag{11}$$

$$R = R_2 - Q \tag{12}$$

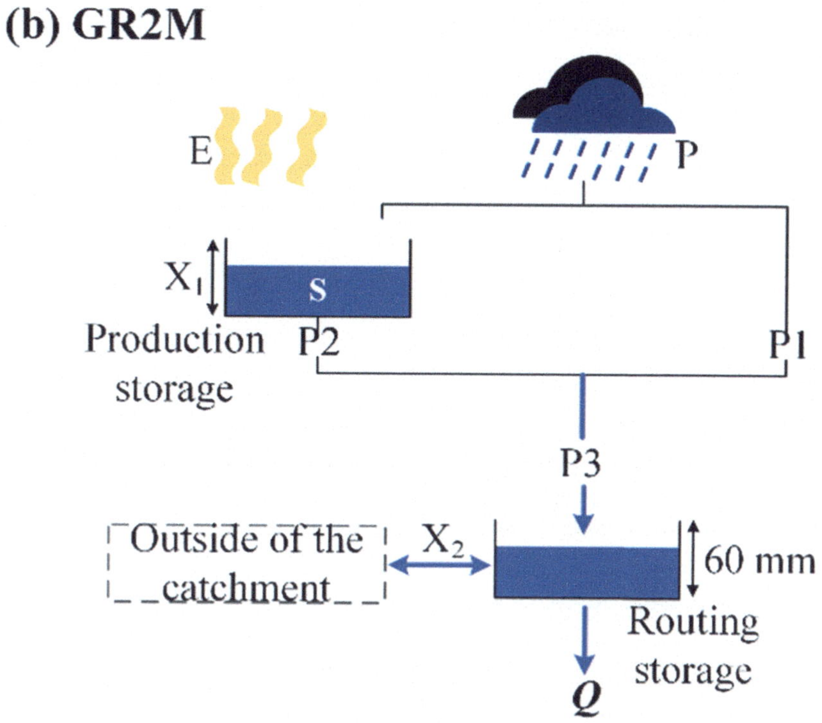

Fig. 6. GR2M precipitation—runoff model [17].

3.12 Determination Coefficient R^2

Expresses the linear dependence between two variables. For the study, it refers to the measured flows and the generated flows.

In the following Table 1, you can see the ranges of R^2.

Table 1. Referential values for the coefficient of determination.

R^2 range	Model quality category
$0.99 \le R^2 < 1.00$	Excellent
$0.95 \le R^2 < 0.99$	Very good
$0.90 \le R^2 < 0.95$	Well
$0.85 \le R^2 < 0.90$	Pretty good
$0.80 \le R^2 < 0.85$	Average
$0.70 \le R^2 < 0.80$	Satisfying
$R^2 < 0.70$	Insufficient

3.13 Nash-Sutcliffe Coefficient

The Nash-Sutcliffe model efficiency (NSE) coefficient is used to evaluate the predictive ability of hydrological models. Is defined as:

$$E = \frac{\sum_{i=1}^{n}(Qsim, i - Qi)^2}{\sum_{i=1}^{n}(Qi - \overline{Q})^2} \quad (13)$$

where:

$Q_{sim,i}$ Simulated flows in m^3/s.
Q_i Observed flows in m^3/s.
$\overline{Q}$ Average observed flows in m^3/s.

And it measures when the variability of the observations is explained by the simulation. If the simulation is perfect, E = 1; If we tried to fit the observations with the average value, then E = 0 [18].

Table 2 shows the reference values of the Nash Sutcliffe criterion.

Table 2. Reference values of the Nash—Sutcliffe criterion.

E	Adjustment
<0.2	Insufficient
0.2–0.4	Satisfying
0.4–0.6	Well
0.6–0.8	Very good
>0.8	Excellent

4 Methodology

The delimitation of the Ica Basin was carried out up to the La Achirana hydrometric station with the ArcGIS software, which was also used to obtain the maps presented in the research.

A code was created in Java (Script) and Python language for Google Earth Engine that allows us to obtain and average daily climatic values of the basin, which will later be averaged into monthly values for the use of the GR2M model. The GR2M conceptual model is used to simulate monthly discharges in gauged and ungauged subbasins from January 1981 to March 2020 [5]. For this investigation, the RAIN4PE precipitation grid product was used and ERA5 was used for temperatures.

Emphasize that the information available on the RAIN4PE grilled product is found every 10 km of separation, just like ERA5.

A restriction of our study lies in the capacity of Google Earth Engine, since when performing extensive calculations on a daily level, it only allows the download of data

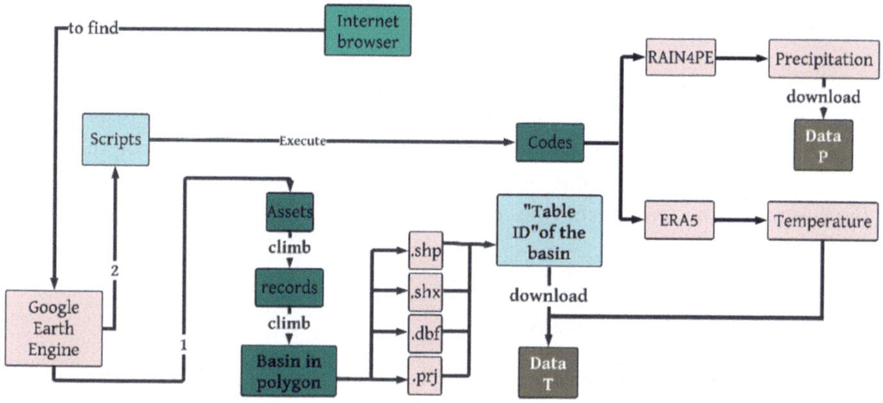

Fig. 7. Flowchart of the steps for data extraction in GEE.

for up to 10 years for the basin area of interest. For this reason, the data acquisition interval was limited to a range of 5–10 years of collection (See Fig. 7).

The codes used in this project are shared below for future use in research:

- ERA5
  https://code.earthengine.google.com/1456743c55ed62ec22949811c75b28c6 [12].
- RAIN4PE
  https://code.earthengine.google.com/d27314726d192f33001ffe0f23c77307 [11].

4.1 Calibration of Model GR2M

The objective of this stage is to identify the values of the model parameters in order to optimally adjust a system as close to the real one that the model represents.

For the calibration of the GR2M method, the flow records extracted from the National Water Authority (ANA) were used in a period of time between 09/01/2013 and 05/01/2015.

4.2 Validation

In this step, it will be determined whether the calculations obtained maintain an acceptable trend with respect to the value of the Nash—Sutcliffe Criterion with the data obtained and simulated in the time period between 09/01/2013 and 12/01/2015.

4.3 Flow Extension

By obtaining our validation positively, we will be able to extract data from Google Earth Engine, from 01/01/1981 to 09/01/2013.

In this way we will obtain data over approximately 30 years to update the flows measured by the ANA Observatory.

5 Results

To develop the GR2M method, the Spreadsheet developed by Charles Perrin and Vazken Andréassian was used [16]. The version shown is presented by Mouelhi [4].

Mouelhi made a coherent chain of global conceptual models of precipitation flow with multiannual, annual, monthly and daily time steps.

Furthermore, the values of the model parameters X1, through a comparative approach.

5.1 Calibration of Model GR2M

For this stage, data obtained from the ANA Observatory with simulated data were used, it was carried out with an area of 2207,278 km^2, and a period of time ranging from 09/01/2013 to 05/01/2015. Table 3 shows the value of the Nash efficiency criterion (Q) obtained is 86.2%.

Table 3. Efficiency criteria (%) obtained in the calibration process.

Efficiency criteria (%)	
Nash (Q)	86.2
Nash (VQ)	65.7
Nash (ln(Q))	36
Bilan	92.3

Table 4 shows the parameters of the GR2M model.

Table 4. GR2M model parameters in the calibration process.

Parameters	Transf.	Current
X_1: Output Res. (mm)	4.46	86.53
X_2: Exchange parameter (mm)	0.8	0.8

Table 5 shows the initial filling values.

Table 5. Initial values for calibration in the Ica River basin at the La Achirana H. S.

Initial values	
Initial fill level S0 (max.: x1 mm)	43.27
Initial fill level R0 (max.: 60 mm)	30

Table 6. Hydrometric data used in the calibration process.

Hydrometric data	
Average observed precipitation (mm/month)	45.27
Average observed ETP (mm/month)	2.00
Average observed flows (mm/month)	31.30
Average roots of observed flows	4.80
Average of the observed log flows	2.80

Table 6 shows the hydrometric data used in the calibration process.

When carrying out the calibration process of the GR2M model, the adjustment parameters were obtained, in which the measured data (National Water Authority) and the monthly flows simulated by the GR2M model were compared. The months compared from 05/01/2013 to 05/01/2015 are shown (See Fig. 8).

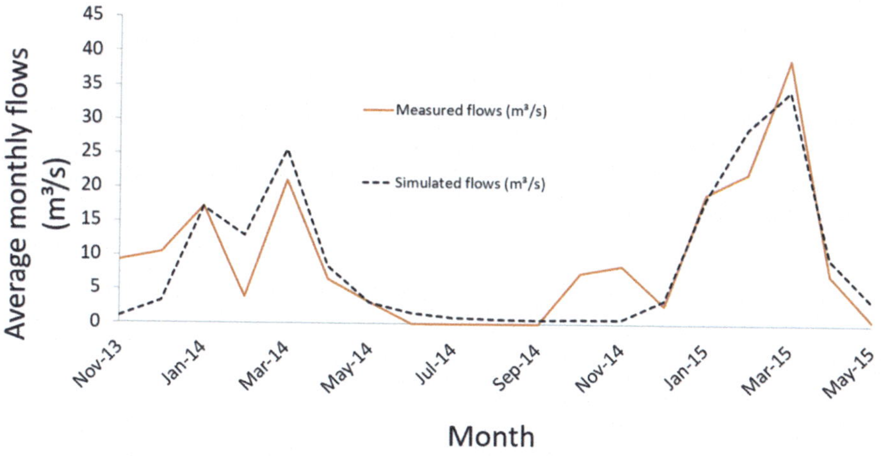

Fig. 8. Comparison of measured monthly flows (ANA) and simulated flows (GR2M) for the Ica Basin in the calibration process between 09/01/2013 until 05/01/2015.

The correlation coefficient (R^2) is shown between the measured daily flows (ANA) and simulated flows (GR2M) for the Ica River basin in the calibration process between 05/01/2013 and 05/01/2015 (See Fig. 9).

Considering a linear trend line, a correlation coefficient (R^2) equal to 0.8864 is obtained.

5.2 Validation

In this analysis stage, the period between 09/01/2013 and 12/01/2015 was designated, containing the calibration stage, to determine if the calculations obtained maintain a similar trend.

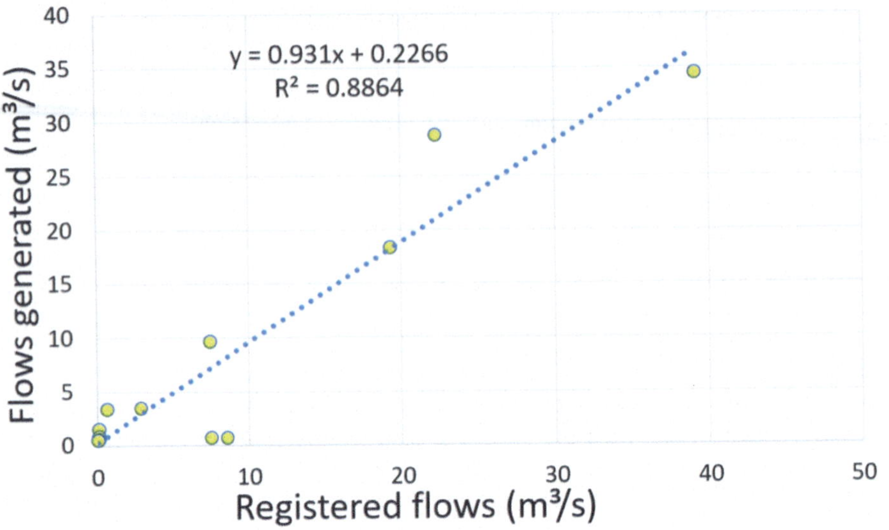

Fig. 9. Correlation of R^2 between measured daily flows (ANA) and simulated flows (GR2M) for the calibration period.

In Table 7, it is observed that the value of the efficiency criterion obtained is 85.2%. Considering the reference values of the Nash-Sutcliffe criterion, the result obtained is in the range of " >0.8". Therefore, it is interpreted that the adjustment made is excellent corresponding to Table 2.

Table 7. Efficiency criteria (%) obtained in the validation process.

Efficiency criteria (%)	
Nash (Q)	85.2
Nash (VQ)	59.9
Nash (ln(Q))	25.1
Bilan	86.7

Table 8 shows the parameters of the GR2M method calculated for the Ica Basin.

Table 8. GR2M model parameters in the validation process.

Parameters	Transf.	Current
X_1: Output Res. (mm)	4.85	128.01
X_2: Exchange parameter (mm)	0.82	0.82

Table 9 shows the initial values for validation.

Table 9. Initial values for the Ica basin in validation.

Initial values	
Initial fill level S0 (max.: x1 mm)	64
Initial fill level R0 (max.: 60 mm)	30

Furthermore, Table 10 shows the hydrometric values obtained using the validation model of the Ica River basin.

Table 10. Hydrometric data used in the validation process.

Hydrometric data	
Average observed precipitation (mm/month)	28.416
Average observed ETP (mm/month)	1.986
Average observed flows (mm/month)	20.728
Average roots of observed flows	3.711
Average of the observed log flows	2.211

The comparison between the measured daily flows (ANA) and simulated flows (GR2M) for the Ica Basin is presented for validation (See Fig. 10).

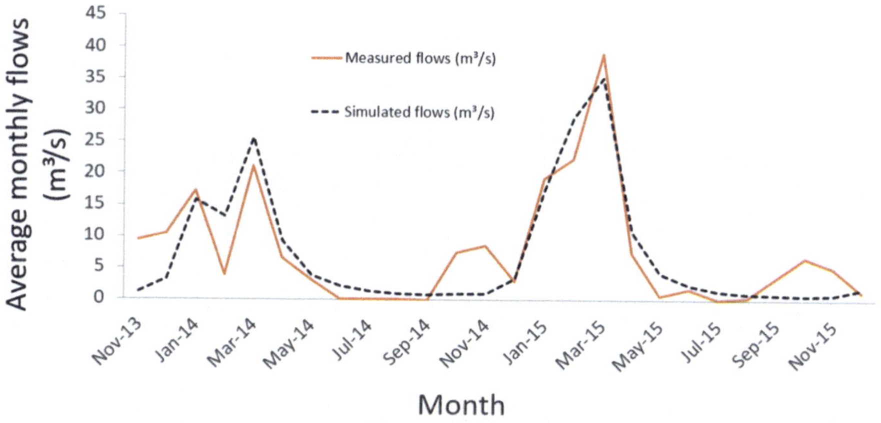

Fig. 10. Comparison of measured daily flows (ANA) and simulated flows (GR2M) for the Ica Basin in the validation process between 09/01/2013 and 12/01/2015.

Figure 11 shows the comparison of estimated and measured flows. Considering a linear trend line, a coefficient of determination (R^2) equal to 0.7966 is obtained.

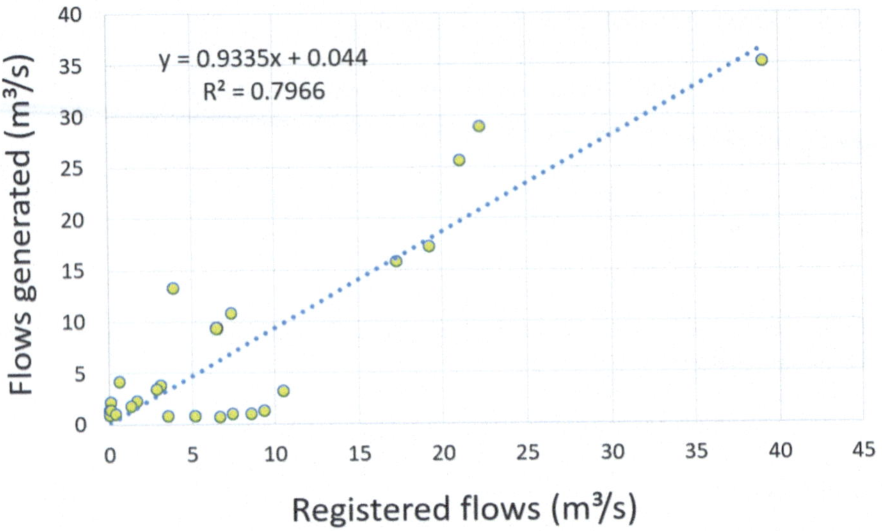

Fig. 11. Correlation of R^2 Comparison of measured daily flows (ANA) and simulated flows (RAIN4PE-ERA5) for the Ica Basin in the validation process between 09/01/2013 and 12/01/2015.

5.3 Flow Extension

Finally, having the calibration and validation parameters as shown in Tables 4 and 10, they were used to obtain the simulated monthly flows between the period from 01/01/1981 to 12/31/2015.

The adjustment between the simulated and measured flows is observed, in turn the simulated years of extension are shown up to the date of 12/31/2015 (See Fig. 12).

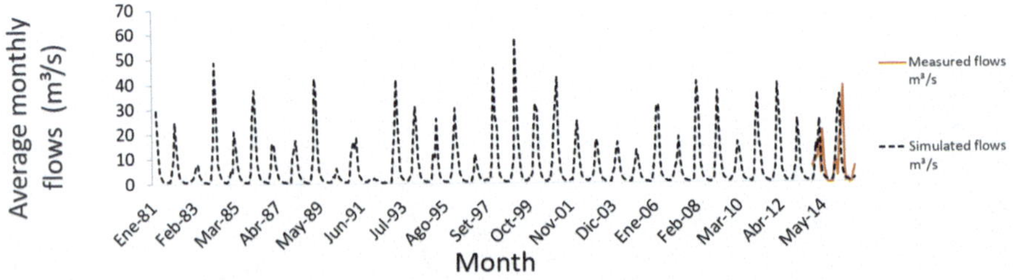

Fig. 12. Daily comparison of measured flows (ANA) and simulated flows (GR2M) for the Ica Basin in the period between 01/01/1981 until 12/31/2015.

These flows were measured by the ANA Observatory and the information from the climate records of precipitation and temperature, obtained from RAIN4PE and ERA5 respectively from the Google Earth Engine platform for data extension.

6 Discussion of Results

The results obtained by applying the GR2M model in the Ica River basin up to the La Achirana hydrometric station are aligned with previous studies that have used the same model in similar hydroclimatic conditions. For example, Makhlouf and Michel demonstrated the effectiveness of the GR2M model to simulate monthly flows in French basins, obtaining Nash-Sutcliffe coefficients comparable to ours [19].

Xiong and Guo also found that GR2M provided accurate simulations in various watersheds [20].

Furthermore, research such as that of Fernández et al. (2009), who applied the model in a forest basin in Spain, showed a similar predictive capacity, validated by significant coefficients of determination (R^2) [21].

On the other hand, recent studies such as those by Oudin et al. (2005) and Perrin et al. (2003) have contributed to improving the model, suggesting adjustments in the potential evapotranspiration inputs to improve the precision of the simulations, which is relevant for future research in our region [22, 23].

This comparison not only validates our findings, but also highlights potential areas for methodological improvements, such as the integration of additional data and the use of improved evapotranspiration inputs.

The findings of this study are significant for the management of water resources in the Ica River basin. The ability of the GR2M model to predict monthly flows with high precision provides a valuable tool for the planning and management of construction works for dams, river defenses, etc. This is particularly relevant given the context of recent climate disasters in the region. Furthermore, the use of high-resolution climate data, such as ERA5 and RAIN4PE, underlines the importance of integrating advanced technologies and accurate data in hydrological modeling.

Despite the positive results, this study has several limitations. Restriction of the data acquisition interval due to the capabilities of Google Earth Engine limited the temporality of the analyzes to periods of 5 to 10 years. Future research could benefit from greater data processing power, allowing for a more extensive and detailed assessment of long-term trends. Furthermore, the accuracy of the results could be improved by incorporating additional hydrometeorological data and using complementary models to validate the findings.

From a practical perspective, the results of this study can be used by local and regional authorities to design and implement more effective disaster mitigation strategies. Theoretically, this work contributes to the body of knowledge on flow modeling in watersheds using the GR2M model and highlights the usefulness of geospatial analysis platforms such as Google Earth Engine.

7 Conclusions

This study has managed to generate a series of monthly flows in the Ica River basin up to the La Achirana hydrometric station using the GR2M reception-runoff model. The main results indicate that the model is capable of simulating flows with high precision, evaluated by the Nash-Sutcliffe coefficient and the coefficient of determination (R^2), which showed values indicative of a good predictive capacity.

According to the analysis of this research, optimal performance of the hydrological model was demonstrated. In the calibration stage, the value of the Nash efficiency criterion was 86.2%; in validation, 85.2%. Considering the range of values, both are within an excellent range. Thus, showing the proper fit of the model.

We obtained similar correlation coefficients (R^2), since these are 0.8864 for calibration and 0.7966 for validation. The value of the Bilan criterion in the first stage is 92.3% and in the next stage it is 86.7%.

The flows simulated in this research will be used for future hydraulic irrigation projects for agricultural lands taking into account the La Achirana Hydrometric Station. However, the study also has limitations, mainly in the data processing capacity and the temporality of the analyses, which could be addressed in future research through the use of more data and advanced computational resources.

In summary, this research not only contributes to the understanding of hydrological dynamics in the Ica River basin, but also offers a basis for future research and practical applications in water resources management. It is suggested that future studies explore the integration of additional models and the expansion of analysis periods to improve the precision and applicability of the results.

References

1. El Comercio, https://elcomercio.pe/peru/ica/desborde-del-rio-ica-dejo-mas-de-100-hectareas-de-cultivo-afectadas-nnpp-noticia/?ref=ecr#google_vignette. Last accessed 04 May 2024
2. LA LUPA, https://lalupa.pe/ica/rio-ica-se-desborda-fuerte-caudal-24-enero-2024-80648/. Last accessed 04 May 2024
3. LA LUPA, https://lalupa.pe/ica/rio-ica-inundacion-en-ocucaje-afecta-viviendas-y-servicios-tras-desborde-de-canal-81783/. Last accessed 04 May 2024
4. Mouelhi, S.: Stepwise development of a two-parameter monthly water balance model. J. Hydrol. **318**(1–4), 200–214 (2006)
5. Llauca, H.: PISCO_HyM_GR2M: a model of monthly water balance in Peru (1981–2020). Water **8**(13), 1048 (2021)
6. Burgan, H.: Monthly flow duration curve model for ungauged river basins. Water **12**(2), 338 (2020)
7. Alfaro, R., Traverso, K., Mamani, J.: Generation of monthly average flow rates from the hydrological characteristics in the huancane river basin. In: Calvo, L. (ed.) Proceedings of the 38th IAHR World Congress, pp. 2791–2800. Panama City, Panama, IAHR, Madrid (2019)
8. ANA, https://snirh.ana.gob.pe/VisorPorCuenca/. Last accessed 04 May 2024
9. ArcGIS Resources, https://resources.arcgis.com/es/help/getting-started/articles/026n00000014000000.htm. Last accessed 04 May 2024
10. Google Earth Engine, https://earthengine.google.com/platform/. Last accessed 04 May 2024
11. Carmona, A.: Google Earth Engine script for time series analysis of satellite images (RAIN4PE), https://code.earthengine.google.com/d27314726d192f33001ffe0f23c77307. Last accessed 04 May 2024
12. Carmona, A.: Google Earth Engine script for time series analysis of satellite images (ERA5), https://code.earthengine.google.com/1456743c55ed62ec22949811c75b28c6. Last accessed 04 May 2024
13. ANA, https://www.ana.gob.pe/contenido/la-autoridad-nacional-del-agua. Last accessed 04 May 2024

14. Fernández, C.: Hydrology session: Tools for water resources management in Peru. PIK Potsdam, https://www.pik-potsdam.de/en/institute/departments/activities/epicc/partner-countries/peru/activities-knowledge-transfer/3_CARLOS_FERNANDEZ_RAIN4PE_sesion_hidrologia_B_EPICC.pdf. Last accessed 04 May 2024
15. Copernicus Climate Data Store, https://cds.climate.copernicus.eu/cdsapp#!/dataset/reanalysis-era5-single-levels?tab=overview. Last accessed 04 May 2024
16. WebGR, https://webgr.inrae.fr/outils/modeles-hydrologiques/modele-hydrologique-mensuel-gr2m. Last accessed 04 May 2024
17. Cheng, L.: Improving the applicability of lumped hydrological models by integrating the generalized complementary relationship. Water Resour. Res. **60**(2), 1–17 (2024)
18. Nash, J.: River flow forecasting through conceptual models part I — A discussion of principles. J. Hydrol. **10**(3), 282–290 (1970)
19. Zoubir, M.: A two-parameter monthly water balance model for French watersheds. J. Hydrol. **162**(3–4), 299–318 (1994)
20. Xiong, L.: A two-parameter monthly water balance model and its application. J. Hydrol. **216**(1–2), 111–123 (1999)
21. Fernández, C.: Streamflow drought time series forecasting: a case study in a small watershed in North West Spain. Stochast. Environ. Res. Risk Assess. **23**, 1063–1070 (2009)
22. Oudin, L.: Which potential evapotranspiration input for a lumped rainfall–runoff model?: Part 2—Towards a simple and efficient potential evapotranspiration model for rainfall–runoff modelling. J. Hydrol. **303**(1–4), 290–306 (2005)
23. Perrin, C.: Improvement of a parsimonious model for streamflow simulation. J. Hydrol. **279**(1–4), 275–289 (2003)

Optimization of Regional Energy-Water Nexus System Under Multiple Uncertainties

Ruoyan Liu[1], Yongping Li[1,2](✉), Guohe Huang[1,2], Yanxiao Zhou[1], and Yanfeng Li[1]

[1] State Key Joint Laboratory of Environmental Simulation and Pollution Control, School of Environment, Beijing Normal University, Beijing, China
yongping.li@iseis.org

[2] Institute for Energy, Environment and Sustainable Communities, University of Regina, Regina, Sask, Canada

Abstract. In this paper, an interval chance-constrained programming (ICCP) method is proposed for collaborative energy-water nexus (EWN) system under multiple uncertainties. This method can solve the uncertainties expressed as probability distribution and interval. An ICCP-based energy-water nexus (ICCP-EWN) model is established, where 15 scenarios are designed for assessing the effect of water cycle coefficients and constraint violation risk. The results show that: (i) coal would remain the main energy sources for future energy supply, but the proportion of clean energy (e.g., natural gas, biomass) in the future power generation structure would significantly increase (about 15.6%); (ii) the water withdrawal of power generation in the energy system would be largest, where thermal power generation would have the largest proportion of water withdrawal, accounting for 85.0%; (iii) the decreasing α level and increasing p level will bring higher system benefits, and the reliability of the system would be reduced. These results can help determine the energy mix of a planned area, improve security of energy supply, and make trade-off between system benefits and the risk of violating constraints.

Keywords: chance-constrained programming · energy-water nexus · Kaikong river · uncertainty

1 Introduction

The rapid economic development and the long-term social stability are inseparable from the continuous development of the energy industry. However, China's energy supply system still has a security crisis. China's conventional energy resources only account for 10.7% of the world's total, per capita energy resources are far lower than the world average [1]. Besides, energy and water are closely related. More than 80% of the world's electricity generation is directly or indirectly dependent on water resources, and a large amount of water is consumed in the process of energy production, processing, and conversion [2, 3]. Therefore, the construction of energy-water nexus systems becomes particularly important for realizing the cooperative management of energy and water resources. In addition, the energy-water nexus system has multiple uncertainties and

G. Huang et al. (Eds.): ICESD 2024, ESE, pp. 70–79, 2025.
https://doi.org/10.1007/978-3-031-88683-6_6

complexities, such as the system parameters related to economic data, energy demand and resource availability. The uncertain parameters and their interactions will affect the accuracy and feasibility of the system management. Therefore, more robust planning methods should be adopted to solve the above problems.

Chance-constrained programming (CCP) is an effective method to solve the stochastic uncertainties in the system. For example, Li et al. (2011) [4] developed a CCP-based optimization method for economic and environmental planning in Tianjin, China, where CCP was applied to treat COD emission capacity. Nikplaos and Pavlos (2019) [5] developed a CCP model to deal with uncertainties in the planning process, which are embodied in load demand and renewable energy generation. Farzaneh and Jalal (2020) [6] presented a CCP-based optimization model for planning centralized power generation expansion where the short-term uncertainty (wind power) is described by a Gaussian distribution. However, there is still uncertain variables represented as interval numbers in EWN system, and CCP method is difficult to deal with the uncertain variables, where the upper and lower bound are known and the specific distribution is unknown. Therefore, an interval linear programming (ILP) method should be introduced to handle this problem [7].

Therefore, this study aims to develop an interval chance-constrained programming (ICCP) to plan EWN systems under uncertainties. ICCP integrates CCP and ILP into a common framework. The contribution includes: (i) ICCP can both resolve uncertainties in the form of interval parameters on the left of the constraints and uncertainties expressed as probability distribution on the right of the constraints; (ii) an ICCP-EWN model is developed for the Kaikong River basin, considering the energy system structure under different water resource availabilities and water cycle coefficients; (iii) the results can analyze the interaction between system economic and water conservation. The result not only helps managers to adjust the current energy mix of the plan area, but also helps managers to understand the tradeoff between system benefits and the risk of violating constraints.

2 Methodology

By integrating interval linear programming (ILP) with chance-constrained programming (CCP), an interval chance-constrained programming method (ICCP) is constructed and used to process the uncertainty information expressed as interval number and probability distribution in EWN system. The ICCP model can be expressed as [8]:

$$Max\, f^{\pm} = C^{\pm}X^{\pm} \tag{1}$$

subject to:

$$A_i^{\pm}X^{\pm} \leq B_i(t)^{(p_i)},\ i = 1, 2, \ldots\ldots, m \tag{2}$$

$$A_i^{\pm} \in A^{\pm},\ b_i(t) \in B(t),\ i = 1, 2, \ldots\ldots, m \tag{3}$$

$$x_j^{\pm} \geq 0,\ x_j^{\pm} \in X^{\pm}, j = 1, 2, \ldots\ldots, n \tag{4}$$

$$p_i \in [0, 1] \tag{5}$$

According to the interactive algorithm of interval parameter planning, the upper and lower bound submodel of interval chance-constrained programming can be obtained.

Upper bound submodel:

$$Max\, f^+ = \sum_{j=1}^{k_1} C_j^+ x_j^+ + \sum_{j=k_1+1}^{n} C_j^+ x_j^- \tag{6}$$

subject to:

$$\sum_{j=1}^{k_1} |a_{ij}|^- sign(a_{ij}^-) x_j^+ + \sum_{j=k_1+1}^{n} |a_{ij}|^+ sign(a_{ij}^+) x_j^- \le b_i(t)^{(p_i)}, \quad \forall i \tag{7}$$

$$x_j^{\pm} \ge 0, \ \forall j \tag{8}$$

Lower bound submodel:

$$Max\, f^- = \sum_{j=1}^{k_1} C_j^- X_j^- + \sum_{j=k_1+1}^{n} C_j^- X_j^+ \tag{9}$$

subject to:

$$\sum_{j=1}^{k_1} |a_{ij}|^+ sign(a_{ij}^+) x_j^- + \sum_{j=k_1+1}^{n} |a_{ij}|^- sign(a_{ij}^-) x_j^+ \le b_i(t)^{(p_i)}, \quad \forall i \tag{10}$$

$$x_j^{\pm} \ge 0, \ \forall j \tag{11}$$

$$x_j^- \le x_{jopt}^+, j = 1, 2, \ldots\ldots k_1 \tag{12}$$

$$x_j^+ \ge_{xjopt}^-, j - k_1 + 1, k_1 + 2, \ldots n \tag{13}$$

By solving the two submodels, the optimal solutions of decision variables and system objectives can be obtained: $x_{jopt}^{\pm} = [x_{jopt}^-, x_{jopt}^+], f_{jopt}^{\pm} = [f_{jopt}^-, f_{jopt}^+]$.

3 Case Study

3.1 Overview of the Study System

The Kaikong River basin is located in Bayingoleng Mongolian Autonomous Prefecture, Xinjiang, China, extending from 82°94′ to 90°54′ east longitude and 39°57′ to 43°36′ north latitude. It consists mainly of Kaidu River, Kongque River, and one lake, Bosten Lake [9]. The Kaikong River Basin faces the scarcity of water resources, the surface water resources in the basin are 4.20 billion m^3. The Kaikong River basin is an important energy production base with abundant fossil resources. The local energy structure is still lead by traditional power generation modes such as thermal power generation, and the shortage of water resources will greatly affect the total power generation [10]. Furthermore, thermal power generation consumes significant amount of water resources and have a negative impact on the ecological environment, exacerbating the issue of water scarcity. The security of energy and water resources is crucial for the economic development and ecological safety in the basin, so it is urgent to manage the energy-water nexus system in a collaborative manner.

3.2 Model Formulation

An ICCP-based EWN management (ICCP-EWN) model was developed for the Kaikong River basin. The goal is to maximize the benefits of the energy system within the planning scope. The model includes six administrative regions in the Kaikong River basin (Korla City, Yuli County, Yanqi County, Hejing County, Heshuo County, Bohu County), twelve energy types (raw coal, crude oil, natural gas, washed coal, coke, gasoline, kerosene, diesel oil, fuel oil, LNG, electricity and heat), three energy processing types (coal washing, coking and oil refining) and eleven energy conversion types (coal-fired power, coal cogeneration, gas-fired power, gas cogeneration, hydropower, wind power, solar power, biomass, coal-fired heating, gas-fired heating, geothermal heating) and seven planning periods (2025–2060, 5 years as a planning period). In addition, to resolve the random uncertainty in the EWN system, we consider five p levels (i.e. p = 0.01, 0.05, 0.10, 0.15, and 0.20). In order to explore the influence of cooling water cycle efficiency on WEN system, the model also involves three water cycle coefficients (low, medium and high), which lead to 15 scenarios. The ICCP-WEN model is expressed as:

$$Maxf^{\pm} = Benefit^{\pm} - Cost^{\pm} \tag{14}$$

$$Benefit^{\pm} = \underbrace{\sum_{i=1}^{6}\sum_{u=1}^{7}\sum_{k=1}^{12}\sum_{t=1}^{7} RED_{k,t}^{\pm} \times FED_{i,u,k,t}^{\pm}}_{\text{Benefit for sale}} + \underbrace{\sum_{i=1}^{6}\sum_{k=1}^{12}\sum_{t=1}^{7} REX_{k,t}^{\pm} \times AEX_{i,k,t}^{\pm}}_{\text{Benefit for energy outport}} \tag{15}$$

$$
\begin{aligned}
Cost^{\pm} = {} & \underbrace{\sum_{i=1}^{6}\sum_{k=1}^{3}\sum_{t=1}^{7} CEE_{k,t}^{\pm} \times AEE_{i,k,t}^{\pm} + \sum_{i=1}^{I}\sum_{m=1}^{3}\sum_{t=1}^{T} CEP_{m,t}^{\pm} \times AEP_{i,m,t}^{\pm}}_{\text{Cost for energy exploitation and processing}} + \underbrace{\sum_{i=1}^{6}\sum_{k=1}^{12}\sum_{t=1}^{7} CEI_{k,t}^{\pm} \times AEI_{i,k,t}^{\pm}}_{\text{Cost for energy import}} \\
& + \underbrace{\sum_{i=1}^{6}\sum_{n=1}^{11}\sum_{t=1}^{7} FCEC_{n,t}^{\pm} \times CEC_{i,n,t}^{\pm} + VCEC_{n,t}^{\pm} \times AEC_{i,n,t}^{\pm} + VCCC_{n,t}^{\pm} \times ECC_{i,n,t}^{\pm}}_{\text{Cost for electricity generation}} \\
& + \underbrace{\sum_{i=1}^{6}\sum_{m=1}^{3}\sum_{t=1}^{7} AEP_{i,m,t}^{\pm} \times UEOP_{m,t}^{\pm} \times ROM_{t}^{\pm} \times COM_{t}^{\pm} + \sum_{i=1}^{6}\sum_{n=1}^{11}\sum_{t=1}^{7} AEC_{i,n,t}^{\pm} \times UEOC_{n,t}^{\pm} \times ROM_{t}^{\pm} \times COM_{t}^{\pm}}_{\text{Cost for carbon reduction}} \\
& + \underbrace{\sum_{i=1}^{6}\sum_{q=1}^{3}\sum_{m=1}^{3}\sum_{t=1}^{7} AEP_{i,m,t}^{\pm} \times UESP_{q,m,t}^{\pm} \times RSM_{q,t}^{\pm} \times CSM_{q,t}^{\pm} + \sum_{i=1}^{6}\sum_{q=1}^{3}\sum_{n=1}^{11}\sum_{t=1}^{7} AEC_{i,n,t} \times UESC_{q,n,t}^{\pm} \times RSM_{q,t}^{\pm} \times CSM_{q,t}^{\pm}}_{\text{Cost for pollutant reduction}} \\
& + \underbrace{\left(\sum_{i=1}^{6}\sum_{k=1}^{12}\sum_{t=1}^{7} AEE_{i,k,t}^{\pm} \times EWEX_{k,t}^{\pm} + \sum_{i=1}^{6}\sum_{m=1}^{3}\sum_{t=1}^{7} AEP_{i,m,t}^{\pm} \times EWPR_{m,t}^{\pm} + \sum_{i=1}^{6}\sum_{n=1}^{11}\sum_{t=1}^{7} AEC_{i,n,t}^{\pm} \times EWC_{n,t}^{\pm} \times \alpha \right) \times CWC_{t}^{\pm}}_{\text{Cost for water consumption}}
\end{aligned}
\tag{16}
$$

subject to:

(1) Energy exploitation constraint

$$AEE_{i,k,t}^{\pm} \leq MAXAEE_{i,k,t}^{\pm}\, k = 1, 2, 3; \forall i, t \tag{17}$$

(2) Restriction on energy transfer in and out

$$AEI_{i,k,t}^{\pm} \leq MAXEI_{i,k,t}^{\pm} \quad \forall i, k, t \tag{18}$$

$$AEX_{i,k,t}^{\pm} \leq MAXEX_{i,k,t}^{\pm} \quad \forall i, k, t \tag{19}$$

(3) Renewable energy constraints

$$AEC_{i,n,t}^{\pm} \leq MAXrenew_{i,n,t}^{\pm} \;\; n = 5, 6, 7, 8; \;\; \forall i, t \tag{20}$$

(4) Water resource availability constraints

$$\Pr\left\{ \begin{array}{l} \sum_{k=1}^{3}\sum_{t=1}^{T} EWEX_{k,t}^{\pm} \times AEE_{i,k.t}^{\pm} + \sum_{m=1}^{3}\sum_{t=1}^{T} EWPR_{m,t}^{\pm} \times AEP_{i,m,t}^{\pm} \\ + \sum_{n=1}^{4}\sum_{t=1}^{T} EWC_{n,t}^{\pm} \times AEC_{i,n,t}^{\pm} \times \alpha + \sum_{i=1}^{I}\sum_{t=1}^{T} AEC_{i,n=5,t}^{\pm} \times EWW_{t}^{\pm} \leq TWE_{i,t} \end{array} \right\} \geq 1 - p \tag{21}$$

(5) Expansion constraint

$$AEP_{i,m,t}^{\pm} \leq RCP_{i,m,t}^{\pm} \quad \forall i, m, t \tag{22}$$

$$CEC_{i,n,t}^{\pm} = CEC_{i,n,t-1}^{\pm} + ECC_{i,n,t}^{\pm} - TY_{i,n,t}^{\pm} \quad t = 2, 3, \ldots, 6; \;\; \forall i, n \tag{23}$$

$$CEC_{i,n,t}^{\pm} \leq CEU_{i,n,t}^{\pm} \quad \forall i, n, t \tag{24}$$

(6) Energy conversion constraint

$$CEC^{\pm}_{i,n,t} \cdot OT^{\pm}_{n,t} \geq AEC^{\pm}_{i,n,t} \quad \forall i, n, t \tag{25}$$

(7) Nonnegative constraints

$$AEC^{\pm}_{i,n,t} \geq 0, \; CEC^{\pm}_{i,n,t} \geq 0, \; ECC^{\pm}_{i,n,t} \geq 0, \; \forall i, n, t; AEE^{\pm}_{i,k,t} \geq 0, \; \forall i, k, t \tag{26}$$

4 Result and Discussion

Figure 1 shows the power generation of different regions under different scenarios. The results show that during the planning period, the power generation in the Kaikong River basin has an upward trend (i.e., an average annual increase of about 15%). The total power generation of Korla city during the planning period would be the highest, which would be 90.9 × 10^3 GWh, accounting for 70.4% of the total power generation in the basin. There are several power plants and cogeneration projects in the city of Korla, and the forecast results would accord with the actual situation. With the increase of default probability p, there is no significant change in power generation. This is because a higher p level means a higher risk of violating constraints and therefore more water allocation can be obtained. Result indicates that increasing the amount of water supplied would not significantly increase its power generation, and other efforts (such as developing clean energy and increasing energy extraction) are needed to increase local power generation.

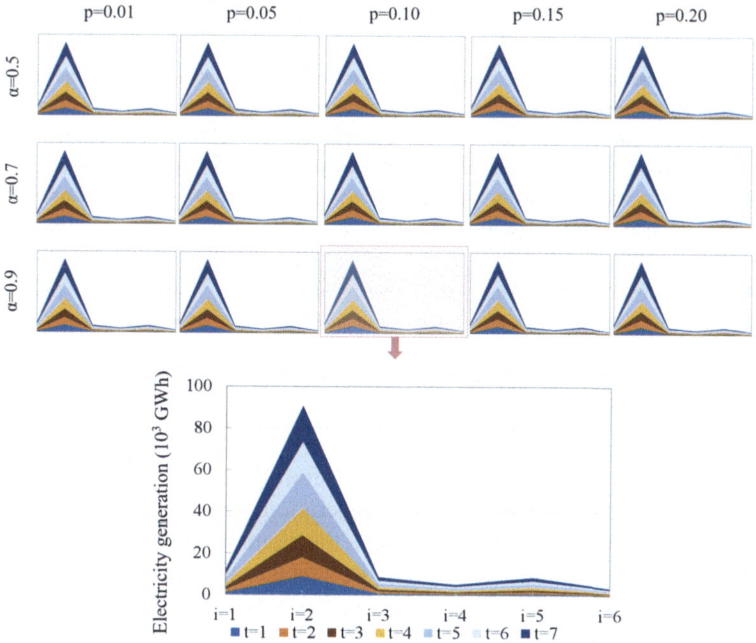

Fig. 1. Electricity generation under different scenarios.

Figure 2 shows the proportion of new unit capacity for different energy conversion types. The results show that the new unit capacity of traditional coal-fired power generation and coal-fired cogeneration accounts for the largest proportion, accounting for 52.2–98.3%. Next are relatively clean gas-fired power generation and gas-fired cogeneration, and the proportion of newly built units for these two types of total energy conversion is increasing during the planning period, reaching a maximum of 45.7%. The proportion of new unit capacity for renewable energy (i.e. hydropower, wind power, thermal power, and biomass power generation) is relatively small, possibly due to the high unit expansion cost. In situations where electricity demand cannot be fully met, the system still needs to prioritize the development of traditional energy conversion types with lower power generation costs.

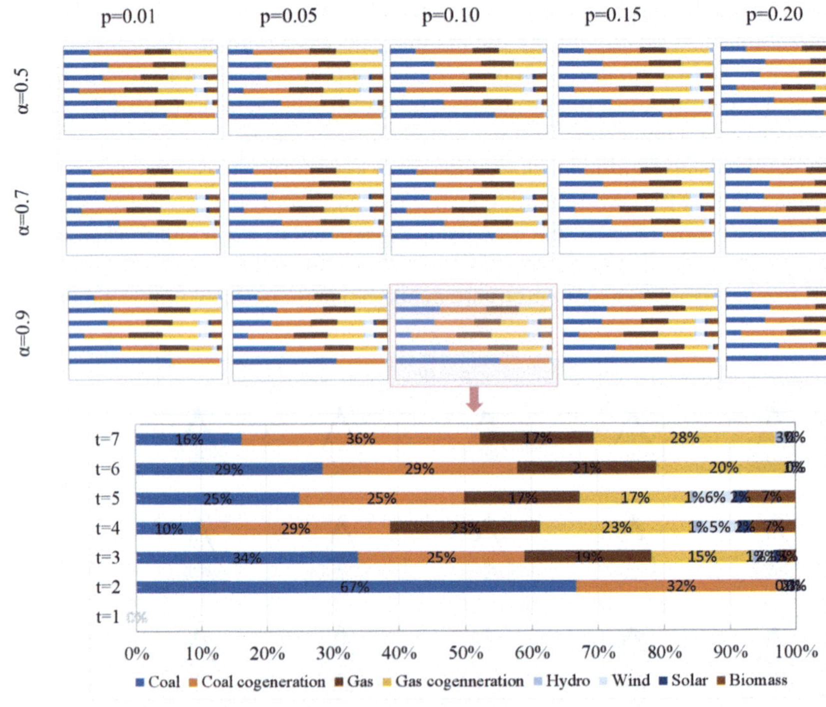

Fig. 2. Proportion of expansion capacity for different energy conversion types.

Figure 3 shows the water withdrawal of each power generation mode of the energy system under different scenarios. As shown, the future energy system water withdrawal would still be mainly use for thermal power generation (about 85.0%), and this proportion will first increase and then decrease during the planning period. Therefore, thermal power generation is the major user of water in the energy system, and the shortage of water resources will seriously affect the local power supply. On the whole, the total water withdrawal of the energy system for power generation increases first and then decreases during the planning period, reaching a peak value in the planning period t = 5 (2046–2050). Compared with the water withdrawal in the period t = 1, the maximum increase is 46.0%, and then the decrease is achieved. The level of regional electricity demand increases correspondingly with the level of development, and the power generation structure in the region is still relatively traditional, so the water withdrawal in the early stage is large. After the development of renewable energy in the region, the power structure has improved, and the water withdrawal has decreased correspondingly. Compared with the scenario with low water cycle coefficient, the scenario with high water cycle coefficient has higher water saving potential, and the total water withdrawal will be reduced by 8.05×106 m^3.

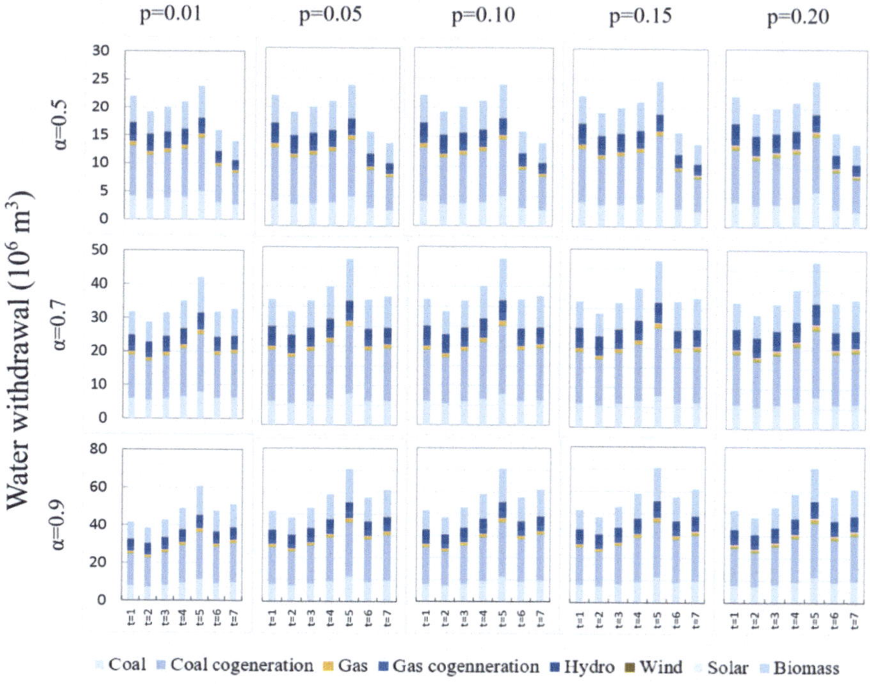

Fig. 3. Water withdrawal of different energy conversion types.

Figure 4 shows the system benefits under different scenarios, indicating that the system benefits will range from 17.9×10^{12} yuan to 58.0×10^{12} yuan. As the level of α increases (i.e., the proportion of water cooling increases), the system efficiency will decrease by up to 0.29%. In addition, as the p level increases (i.e., from 0.01 to 0.20), the system benefit will increase by 1.20–1.65%. Therefore, as the α value decreases and the p value increases, the system benefit would increase. Compared to scenarios with a higher risk of constraint violation (i.e., a larger p value), the system benefit in scenarios with a lower risk will decrease by 0.26×10^{12}–0.73×10^{12} yuan. In order to ensure water security in the river basin, the system needs to sacrifice a certain amount of benefit. Moreover, decisions regarding energy and water require a comprehensive understanding of each other's needs. It is suggested to adjust the energy supply structure and develop towards the renewable energy structure. Financial policies can be adopted to encourage the development of wind and solar photovoltaic power generation.

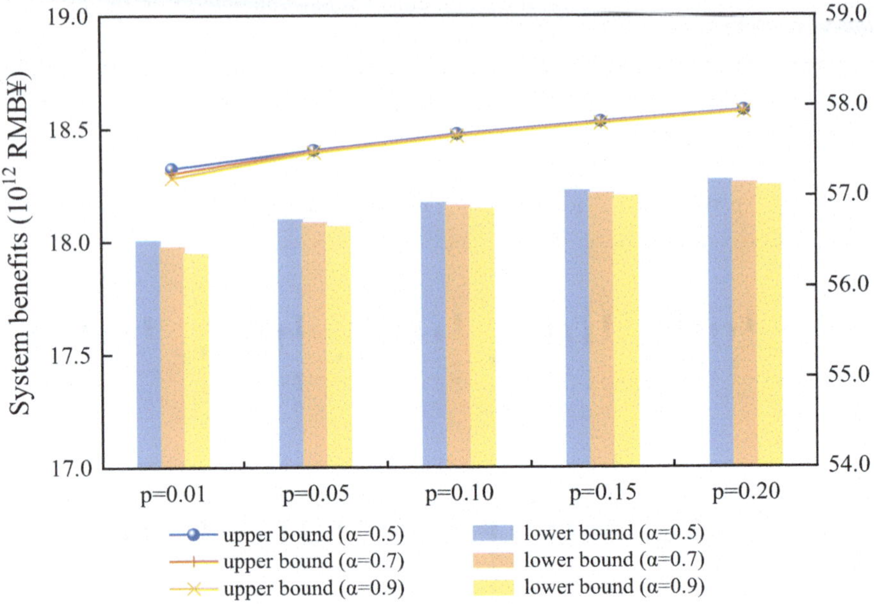

Fig. 4. System benefits under all scenarios.

5 Conclulsion

An interval chance-constrained programming (ICCP) is proposed to cooperatively plan energy-water nexus (EWN) systems under multiple uncertainties. ICCP is capable of dealing with uncertainties expressed as interval numbers and probability distributions. An ICCP-EWN model was established for the Kaikong River basin. Five p levels and three water cycle coefficients (low, medium and high) were considered, which resulted in 15 scenarios. The research results can assist managers in adjusting the local energy structure. Some of the key findings are: (i) coal is still one of the main energy sources of future energy supply, but clean energy (such as natural gas and biomass power generation) will also significantly increase in the proportion of future power generation structure (up 15.6%); (ii) thermal power generation is the largest water user in the energy system, accounting for 85.0% of the total water withdrawal. The local power generation still relies primarily on traditional thermal power generation, thus water shortages can seriously affect local power supplies; (iii) Decreasing the α level (from 0.9 to 0.5) or increasing the p level (from 0.01 to 0.20) would both lead to higher system benefits, which would also decrease the system's reliability. Therefore, EWN system planning under uncertainties should be considered, and the reasonable planning schemes would be obtained by balancing the system benefits and the risk of violating constraints.

However, there are also some limitations and potential areas of future research that should be considered: (i) climate change brings new challenges, and can greatly influence the availability of water resources. Energy-water nexus system needs a transition towards to a more sustainable energy mix, which would lead the energy-water nexus to more complex; (ii) the selection of cooling technology can significantly impact water withdrawal in power generation, particularly in thermal power plants, which should be considered in future work.

Acknowledgements. This research was supported by the National Natural Science Foundation of China (52279003), the National Key R&D Program of China (2023YFC3206503), and Innovative Research Group of the National Natural Science Foundation of China (52221003).

References

1. BP Statistical Review of World Energy 2020, https://www.bp.com/content/dam/bp/business-sites/en/global/corporate/pdfs/energy-economics/statistical-review/bp-stats-review-2020-full-report.pdf. Last accessed 08 June 2020
2. Gong, J.W., Li, Y.P., Lv, J., et al.: Development of an integrated bi-level model for China's multi-regional energy system planning under uncertainty. Appl. Energy **308**, 118299 (2022)
3. Lv, J., et al.: Planning energy-water nexus system under multiple uncertainties—a case study of Hebei province. Appl. Energy **229**, 389–403 (2018)
4. Li, Y.P., et al.: Optimization of regional economic and environmental systems under fuzzy and random uncertainties. J. Environ. Manag. **92**(8), 2010–2020 (2011)
5. Koutsoukis, N., Georgilakis, P.: A chance-constrained multistage planning method for active distribution networks. Energies **12**(21), 41–54 (2019)
6. Pourahmadi, F., et al.: Distributionally robust chance-constrained generation expansion planning. IEEE Trans. Power Syst. Publ. Power Eng. Soc. **35**(4), 2888–2903 (2020)
7. Li, Y.P., Huang, G.H.: An interval-based possibilistic programming method for waste management with cost minimization and environmental-impact abatement under uncertainty. Sci. Total. Environ. **408**(20), 4296–4308 (2010)
8. Liang, M.S., et al.: Development of non-deterministic energy-water-carbon nexus planning model: a case study of Shanghai, China. Energy (Oxford) **246**, 123300 (2022)
9. Li, W.H., Chen, Z.S., Li, B.F., et al.: Analysis of water demand and stability for oasis in Kaidu-Kongque River Basin, Southern Xinjiang. J. Glaciol. Geocryology **34**(6), 1470–1477 (2012)
10. Jin, S.W., Li, Y.P., Huang, G.H., et al.: Development of an integrated optimization method for analyzing effect of energy conversion efficiency under uncertainty—a case study of Bayingolin Mongol Autonomous Prefecture, China. Energy Convers. Manag **106**, 687–702 (2015)

Global Road Expansion Threat to Biodiversity May Be Underestimated

Denis Vasiliev(✉)

Turiba University, Riga, Latvia
Deniss.Vasiljevs@turiba.lv

Abstract. Road infrastructure is expected to expand globally in the coming decades. The infrastructure construction is expected to be particularly intense in the Global South, where most of the world's biodiversity is concentrated. Current data on road effects on the environment predominantly comes from studies on certain species. This may signal potential underestimation of the environmental impacts, which is likely to result in application of insufficient mitigation measures, ultimately undermining biodiversity conservation efforts. This can pose a global threat to provision of vital ecosystem services, ecological integrity and human wellbeing. Ultimately, the damage caused by road infrastructure development may outweigh economical benefits. Thus, lack of understanding of potential consequences of road infrastructure development is very dangerous and needs to be urgently addressed. In the present article the potential impacts of road infrastructure on biodiversity are reviewed. Based on this review effectiveness of currently widespread mitigation measures is critically evaluated and recommendations to policymakers and conservation practitioners are provided.

Keywords: Roadkill · Sustainable Development · Wildlife Crossings

1 Introduction

Road infrastructure development is currently booming with hundreds of millions of kilometers of roads projected to be constructed in upcoming years [1]. Most of the infrastructure development happens in the Global South, funded through major global and regional programs, including Build Back Better World (B3W) initiative, Belt and Road initiative (BRI), Program for Infrastructure Development in Africa (PIDA) and Initiative for the Integration of the Regional Infrastructure of South America (IIRSA). These programs leverage substantial funding (e.g., under BRI investment of 8 trillion US dollars is projected to be invested by 2049 [2], while under B3W 40 + trillion US dollar investment was pledged for infrastructure development [3]). At the same time, it is well established that road construction often is associated with biodiversity loss and environmental degradation [4]. Key drivers of biodiversity loss associated with the linear infrastructure are considered to be so called "roadkill", collisions of animals with vehicles, and habitat fragmentation [5]. Mitigation measures that are usually applied to mitigate these negative effects on biodiversity are construction of fencing and wildlife

G. Huang et al. (Eds.): ICESD 2024, ESE, pp. 80–89, 2025.
https://doi.org/10.1007/978-3-031-88683-6_7

crossings [6]. However, evidence on effectiveness of these measures predominantly comes from studies on individual species rather than biodiversity in general and from the Global North rather than from tropics [6]. It is therefore unclear whether application of the traditional mitigation measures alone is sufficient for protecting biodiversity. If this is not the case, road infrastructure expansion across the tropics is likely to have devastating consequences, as mitigation measures simply may not be sufficiently effective [6, 7]. This is particularly dangerous as potential effects of road infrastructure on biodiversity may simply be underestimated, causing unintentional devastation. Thus, the aim of this paper is to review evidence that indicates potential scale of biodiversity loss that may be caused by road infrastructure development, critically evaluate potential of currently widespread mitigation measures to address this threat and to provide recommendations to relevant policymakers and practitioners.

2 Potential Damage to Biodiversity

Roadkill is the most evident direct threat to biodiversity associated with terrestrial linear infrastructure [8]. Often, when roadkill mitigation measures are considered, a particular attention is paid to large-bodied terrestrial vertebrate species, such as large carnivores [9] or ungulates [10]. This is logical, because vehicle collisions with large-bodied animals represents a safety concern to drivers and simultaneously because such species usually play important roles in ecosystem functioning [11]. Mitigation measures are also often designed to reduce roadkill of species that are of regional concern or have special protected status, regardless of their morphology or taxonomy [12]. Focus on roadkill of certain species, however, may lead to significant underestimation of scale of biodiversity loss associated with linear terrestrial infrastructure. Empirical evidence suggests that both vertebrate and invertebrate animal species from across major taxonomic groups, including mammals [13], birds [14], reptiles [15], amphibians [16], mollusks [17], worms [18] and arthropods [19] suffer from roadkill. Although species tend to be killed at different rates a spectrum of the traits making species inherently more vulnerable to collisions with vehicles [13, 20] seems to be much broader than is usually addressed by roadkill mitigation measures.

Body mass is often suggested to play an important role in species vulnerability to collisions with vehicles [21]; however, the relationship between this trait and roadkill rates does not seem to be as straightforward as it may appear at first glance. As was already mentioned, when roadkill mitigation measures are considered, large-bodied species tend to attract particular attention. Indeed, body mass was found to strongly correlate with home range area [22], which means that large-bodied animals generally may be more likely to cross the roads than species with smaller home ranges. However, several comprehensive studies on North American mammals [21, 23] found that highest roadkill rates were observed among medium sized species. Furthermore, a recent pan-European study [13] found that birds and mammals with small body size were more likely to experience road-associated mortality. Indeed, smaller animals may be particularly vulnerable to roadkill as they tend to be less noticeable on roads [24] and smaller brain size may result in lower spatial learning ability [25]. Lower roadkill rates in large-bodied species, however, do not necessarily mean lower vulnerability to roadkill, as species population densities within the landscape and species reproduction rates generally tend to decrease with increasing body size [26]. Thus, it seems highly likely, that roadkill threat may be high for both large-bodied and medium to small bodied species, as the former may suffer severe population declines even with a low rate of roadkill, while the latter evidently suffer from high roadkill rates.

Another trait that merits particular attention is locomotion mode, as it heavily affects roadkill mitigation measure design considerations [27]. It would be very natural to think that species with terrestrial locomotion would be particularly vulnerable to road-associated mortality. However, despite being poorly, if at all, addressed by roadkill mitigation strategies, species with aerial dispersal tend to be highly vulnerable to road-kill. One such group of species, which to our best knowledge is completely ignored includes wind-dispersed biota [28]. This group may be particularly prone to collisions with vehicles, as the dispersers have little to no control over their movement direction. Research is urgently needed to determine roadkill rates within the group. Another much better explored group includes actively flying, guiding, and soaring species. Rates of roadkill in these groups are astonishing. Worldwide hundreds of millions of birds and billions of flying insects are killed annually due to collisions with vehicles [29, 30].

Birds may collide with vehicles either when sitting or walking on or along roads, or when flying across roads [30]. It is often suggested that non- or low-flying birds, ground nesting or ground feeding species and scavengers are particularly vulnerable to collisions with vehicles [31]. However, empirical evidence indicates that birds that fly across roads, sometimes even at high elevations, may be highly vulnerable to road mortality. Lin [32] found that birds have difficulties to avoid vehicles when flying across roads. In a meta-analysis [33] found that highly mobile bird species with large habitat areas tend to be particularly susceptible to roadkill, which may be associated with specifics of their vision as many highly mobile bird species have blind areas in the direction of travel [34]. Also, birds were found to collide with powerlines located above roads, when changing elevation in order to fly above vehicles [35]. Migratory bird species that rely on starlight navigation suffer from collisions with vehicles due to disorientation caused by light pollution from highway lightning and vehicles [36]. The fact that birds may experience road mortality, not only on road surface, but also when flying across roads has important implications for conservation measure design.

For numerous species with a broad range of traits roads act as an “ecological trap”, which is an inappropriate habitat chosen by mistake and leading to the reduction or extinction of species populations [37]. For instance, cool-blooded animal species were found to be attracted to asphalt road surfaces, because these surfaces radiate heat at night (e.g. [38]). Also, a recent study by Hennigar and colleagues [39] revealed that birds are attracted to roads by traffic noise during breeding season, exposing themselves to increased risk of collisions with vehicles and other negative road effects that may undermine population fitness. Ground nesting birds and turtles lay eggs in road verges, making them increasingly exposed to road mortality risk, too (e.g., [40, 41]). Roads tend to be an attractive food source for bird and animal species from across the food-web, including carnivores [9, 42], insectivores [43], omnivores [20], scavengers [44, 45] and even frugivores (especially in tropics) [46], exposing them to road mortality risk.

As follows from above discussion diverse range of traits make species from a broad range of taxonomic groups vulnerable to roadkill (Table 1).

Table 1. Taxonomic groups of species which may be particularly vulnerable to roadkill.

Taxonomic group	Traits increasing roadkill vulnerability
Large mammals (e.g., ungulates, carnivores)	Large body mass, large home ranges
Small mammals	Small brain size
Birds and flying insects	Aerial locomotion, high mobility, starlight navigation
Ground-nesting species	Nesting and feeding on road verges
Cool-blooded animals (e.g., reptiles, amphibians)	Attraction to heat-radiating asphalt
Wind-dispersed biota	Aerial dispersal, lack of control over movement
Scavengers, omnivores, and frugivores	Attraction to food resources on or near roads

A range of other traits may contribute to species vulnerability to roadkill, including flight initiation distance, maximum sprint speed, time of activity (nocturnal, diurnal or crepuscular), longevity, maturity, duration of parental care, sociality, morphology specifics etc. [44], indicating that much broader range of species are likely to be susceptible to road mortality than is usually addressed in mitigation measures. Furthermore, roadkill rates tend increase in locations with higher local abundances [47] and higher species diversity [9, 13], which suggests that in tropics, where biodiversity is very high, road mortality is likely to be particularly high.

Road infrastructure fosters biodiversity decline not only through direct road mortality [48, 49]. Many species of small [50] and large mammals [51], birds [52], herpetofauna [53] and insects [54] were fund to avoid roads, and therefore may be considered relatively immune to roadkill. However, for such species roads act as complete or partial barriers to movement, which may result in habitat isolation [50]. Habitat isolation in turn involves multiple negative ecological effects, including species deprival from essential resources, habitat overexploitation due to increasing local population densities, disruption of genetic integrity and increased vulnerability to environmental stochasticity in isolated populations, that often result in local extinctions and may undermine species long-term persistence [55]. Furthermore, species not suffering directly from habitat isolation may also be affected through link disruption in food webs that may result in cascading extinctions and even community collapse [56]. Indeed, apex predators are often found to avoid roads (e.g. [51, 57]). It is well established that isolation or loss of large carnivores usually leads to trophic cascades—chain of negative effects moving down through lower levels of the food web [58]. Fragmentation by road networks resulting in loss of landscape permeability for many species may be particularly detrimental in tropics, where in near future climate conditions may shift to an extent not experienced on

Earth for millions of years [59], preventing many species from adapting to new environmental conditions through range shifts [60]. Along with habitat isolation, negative edge effects, and habitat loss associated with landscape fragmentation are likely to disrupt key ecosystem functions further reducing species capability for long-term persistence [55].

Road infrastructure may also impair species long-term persistence by reducing population fitness [61]. Traffic noise disrupts wildlife sound communication, which determines pairing and reproductive success for some animal and particularly bird species [62]. Effects of sound communication disruption may be particularly detrimental in tropics, where due to high biodiversity signal nuances play important role in species communication [63]. Noise and light pollution may also indirectly reduce population fitness by causing continuous stress and/or abnormal behavioral changes [63–65]. In addition to that chemicals and dust from roads were found to contaminate both terrestrial and aquatic ecosystems, thereby causing negative effects on ecosystem functioning, wildlife persistence, health and fitness [61, 66].

Large road infrastructure development projects may open human access to otherwise inaccessible wilderness areas [67]. This may be particularly detrimental or biodiversity, especially in the Global South where in many countries land-use zoning is unclear and the rule of law are limited [49]. Human access is usually associated with habitat conversion and fragmentation, fires, resource extraction (both legal and illegal), including poaching, logging and mining [68]. Scale of road infrastructure expansion and its negative effects on biodiversity may be further exacerbated by emergence of illegal roads steaming from legal ones, as has already happened in Brazilian Amazon where three kilometers of the former were created for each kilometer of the latter [69]. As follows from all discussed above, road infrastructure construction in the Global South poses extreme risks to biodiversity (Fig. 1).

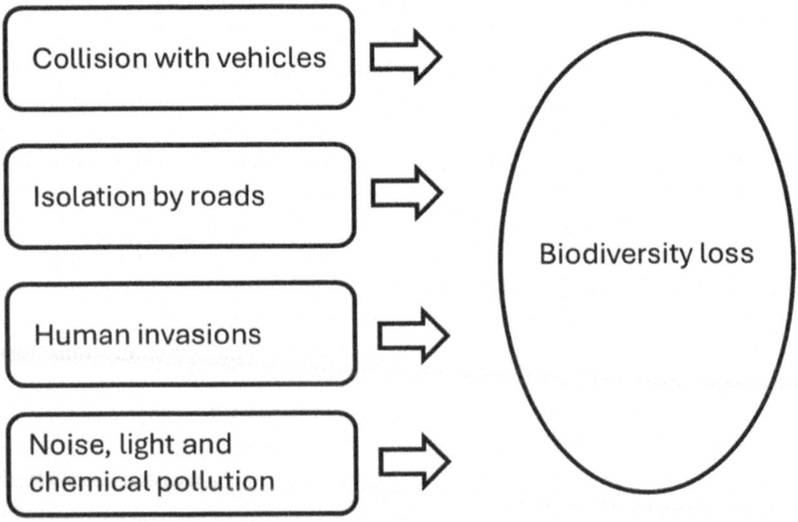

Fig. 1: Kay threats to biodiversity driven by road infrastructure.

3 Mitigation Measures

Currently the most popular measures for mitigating roadkill and habitat fragmentation associated with roads include different types of fencing and crossing structures [6]. Application of these structures was found to be particularly effective when conservation strategies are targeting particular species [6]. However, as follows from the above discussion, these measures are unlikely to be effective in protecting diverse species assemblages from negative road impacts. Another set of measures is based on influencing driver behavior (e.g., signage or seed imitations). These measures, however, tend to have limited effectiveness, especially in remote areas where law enforcement may be problematic. Therefore, alternative strategies need to be considered in areas where biodiversity is particularly high [7].

One of the most effective strategies doubtlessly is road planning that implies avoiding road penetration into highly biodiverse ecosystem, such as tropical forests savannahs, mountainous ecosystems, wetlands etc. Although this option is often considered challenging, methodologies exist that can help achieving this goal without necessarily compromising economic goals [70]. However, it has to be admitted that in cases when there is no key stakeholder support for such solutions, or due to objective reasons it is impossible to plan the road in such a way that it avoids bisecting or closely passing by areas with high biodiversity levels, other solutions need to be considered.

Among other poising strategies is the combination of transport modes [71]. Indeed, when transport infrastructure is planned in a strategic manner it may be possible to combine different transport modes, including land, water and air transport to minimize disturbance to highly biodiverse ecosystems and avoid bisecting them with roads or rail.

When this option is not feasible it is possible to consider construction of tunnels, covered ways or elevated highways [7]. Although these options are often ignored due to high initial investment costs, considering both biodiversity and non-biodiversity benefits in long term can provide a solid justification for these solutions [7]. Indeed, tunnel, elevated highway and/or covered way construction not only can preserve biodiversity and associated ecosystem services, which can have the value of billions of dollars annually, but also the infrastructure itself has proven to have longer lifespan than asphalt, concreate roads or railway [7]. Furthermore, the approaches can help avoiding problems which are often associated with road exploitation in tropics, such as road flooding and road cover overheating and melting [7].

Putting the above constraints together, it is essential to approach road infrastructure planning in a strategic way that integrates biodiversity conservation considerations. Weighing feasibility of different options (Fig. 2), such as completely avoiding road penetration into highly biodiverse ecosystems through road planning or combination of transport modes, or application of modern structures such as tunnels, covered ways or elevated highways need to be considered.

Careful evaluation of the options can allow maximizing economic benefits without compromising economic goals. It is important to remember that reliance only on widespread mitigation techniques may be not sufficient for mitigating negative impacts of road infrastructure on biodiversity, which likely to be associated with ecosystem service loss and associated negative effects on economy and human wellbeing.

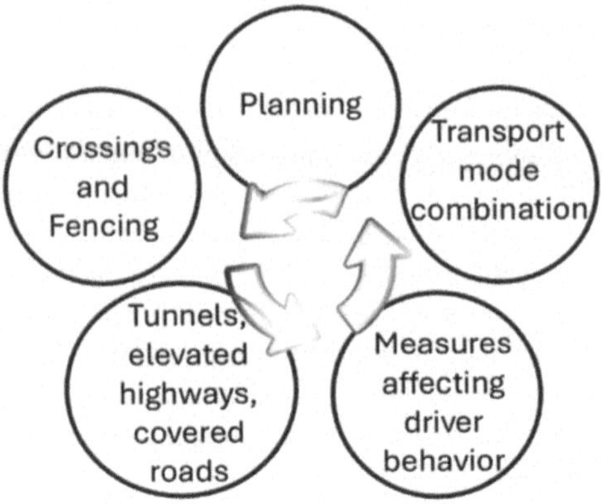

Fig. 2: Integrated approach to road effect mitigation.

4 Conclusions

The ai of this paper was to review evidence that indicates potential scale of biodiversity loss that may be caused by road infrastructure development, critically evaluate potential of currently widespread mitigation measures to address this threat and to provide recommendations to relevant policymakers and practitioners. The literature synthesis revealed that biodiversity impacts of road infrastructure are likely to be significant, especially in the tropical ecosystems where species assemblages are highly biodiverse. Application of the widespread mitigation measures alone is unlikely to be sufficient for mitigation of these effects. As a result, road expansion into biodiverse ecosystems, most of which are located in the Global South, can have devastating consequences. To avoid these consequences it is recommended to consider radical approaches to mitigation of road infrastructure impacts on biodiversity, such as planning roads to avoid highly biodiverse areas, combining transport modes or constructing tunnels, covered ways or elevated highways. Despite apparent complexity and high initial investment costs application of the radial strategies is likely to have long term benefits that are likely to cover associated costs. For the proposed approaches to be implemented with maximal effectiveness it is essential to integrate biodiversity conservation into strategic infrastructure planning policies and practice.

References

1. Meijer, J., et al.: Global patterns of current and future infrastructure. Environ. Res. Lett. **13**(6), 1–10 (2018)
2. Huang, Y.: Understanding China's belt and road initiative: motivation, framework and assessment. China Econ. Rev. **40**, 314–321 (2016)
3. The White House: FACT SHEET: President Biden and G7 Leaders Launch Build Back Better World (B3W) Partnership, https://www.whitehouse.gov/briefing-room/statements-releases/2021/06/12/fact-sheet-president-biden-and-g7-leaders-launch-build-back-better-world-b3w-partnership/. Last accessed 7 May 2024

4. Torres, A., Jaeger, J.A.G., Alonso, J.C.: Assessing large-scale wildlife responses to human infrastructure development. Proc. Natl. Acad. Sci. **113**, 8472–8477 (2016)
5. Laurence, W.F., Goosem, M., Laurance, S.G.W.: Impacts of roads and linear clearings on tropical forests. Trends Ecol. Evol. **24**(12), 659–669 (2009). https://doi.org/10.1016/j.tree.2009.06.009
6. Rytwinski, T., et al.: How effective is mitigation at reducing road kill? a meta-analysis. PloS One **11**, e0166941 (2016)
7. Vasiliev, D., et al.: Light at the end of the tunnel: innovative opportunities for saving tropical biodiversity. Ambio **53**, 702–717 (2024)
8. Forman, R.T.T., Alexander, L.E.: Roads and their major ecological effects. Annu. Rev. Ecol. Evol. Syst. **29**, 207–231 (1998)
9. Planillo, A., et al.: Carnivore abundance near motorways related to prey and roadkills. J. Wildl. Manag. **82**(2), 319–327 (2018)
10. Bíl, M., Kubeček, J., Andrášik, R.: Ungulate-vehicle collision risk and traffic volume on roads. Eur. J. Wildl. Res. **66**(4), 1 (2020). https://doi.org/10.1007/s10344-020-01397-8
11. Tryjanowski, P., et al.: On the origin of species on road warning signs: a global perspective. Global Ecol. Conserv. **27**, e01600 (2021)
12. Collinson, W.J., Marneweck, C., Davies-Mostert, H.T.: Protecting the protected: reducing wildlife roadkill in protected areas. Anim. Conserv. **22**, 396–403 (2019)
13. Grilo, C., et al.: Roadkill risk and population vulnerability in European birds and mammals. Front. Ecol. Environ. **18**(6), 323–328 (2020)
14. Loss, S.R., Will, T., Marra, P.P.: Estimation of bird-vehicle collision mortality on U.S. roads. J. Wildl. Manag. **78**(5), 763–771 (2014)
15. Wagne, R.B., et al.: Snakes on a lane: road type and edge habitat predict hotspots of snake road mortality. J. Nat. Conserv. **61**, 125978 (2021)
16. Puky, M.: Amphibian road kills: a global perspective. In: Proceedings of the 2005 International Conference on Ecology and Transportation (2005)
17. Jeganathan, P., et al.: Seasonal variation in wildlife roadkills in plantations and tropical rainforest in the Anamalai Hills, Western Ghats, India. Curr. Sci. **114**(3), 619–626 (2018)
18. Tabor, R.: Earthworms, crows, vibrations and motorways. New Sci. **62**, 482–483 (1974)
19. Martin, A.E., et al.: Flying insect abundance declines with increasing road traffic. Insect Conserv. Divers. **11**, 608–613 (2018)
20. Cook, T.C., Blumstein, D.T.: The omnivore's dilemma: diet explains variation in vulnerability to vehicle collision mortality. Biol. Cons. **167**, 310–317 (2013)
21. Barthelmess, E.L., Brooks, M.S.: The influence of body-size and diet on roadkill trends in mammals. Biodivers. Conserv. **19**, 1611–1629 (2010)
22. Lindstedt, S.L., Miller, B.J., Buskirk, S.W.: Home range, time, and body size in mammals. Ecology **67**(2), 413–418 (1986)
23. Ford, A.D., Fahrig, L.: Diet and body size of North American mammal road mortalities. Transp. Res. Part D: Transp. Environ. **12**(7), 498–505 (2007)
24. Wright, P.G.R., e al.: Predicting hedgehog mortality risks on British roads using habitat suitability modelling. PeerJ **7**, e8154 (2020)
25. Sherry, D.F., Jacobs, L.F., Gaulin, S.J.C.: Spatial memory and adaptive specialization of the hippocampus. Trends Neurosci. **8**, 298–303 (1992)
26. Damuth, J.: Population density and body size in mammals. Nature **290**, 699–700 (1981)
27. Caceres, N.C.: Biological characteristics influence mammal road kill in an Atlantic Forest-Cerrado interface in south-western Brazil. J. Italian Zool. **78**(3), 379–389 (2011)
28. McLeod, A.M., Leroux, S.J.: Incongruent drivers of network, species and interaction persistence in food webs. Oikos **130**(10), 1726–1738 (2010)
29. Baxter-Gilbert, J.H., et al.: Road mortality potentially responsible for billions of pollinating insect deaths annually. J. Insect Conserv. **19**, 1029–1035 (2015)

30. Husby, M.: Factors affecting road mortality in birds. Ornis Fennica **93**, 212–224 (2016)
31. Glista, D.J., et al.: A review of mitigation measures for reducing wildlife mortality on roadways. Landsc. Urban Plan. **9**, 1–7 (2006)
32. Lin, S.-C.: Landscape and traffic factors affecting animal road mortality. J. Environ. Eng. Landsc. Manag. **24**(01), 10–20 (2016)
33. Rytwinski, T., Fahrig, L.: Do species life history traits explain population responses to roads? a meta-analysis. Biol. Conserv. **147**, 87–98 (2012)
34. Martin, G.R., Shaw, J.M.: Bird collisions with power lines: failing to see the way ahead? Biol. Conserv. **143**, 2695–2702 (2010)
35. Bernardino, J., et al.: Bird collisions with power lines: state of the art and priority areas for research. Biol. Cons. **222**, 1–13 (2018)
36. Ogden, L., Evans J.: Collision course: the hazards of lighted structures and windows to migrating birds. Report published by World Wildlife Fund Canada and Fatal Light Awareness Program (1996)
37. Hale, R., Swearer, S.E.: (2016) Ecological traps: current evidence and future directions. Proc. R. Soc. B **283**, 20152647 (2016)
38. Sullivan, B.K.: Observed differences in body temperature and associated behaviour of four snake species. J. Herpetol. **15**, 245–246 (1981)
39. Hennigar, B., Ethier, J.P., Wilson, D.R.: Experimental traffic noise attracts birds during the breeding season. Behav. Ecol. **30**(6), 1591–1601 (2019)
40. Meunier, F.D., Verheyden, C., Jouventi, P.: Bird communities of highway verges: influence of adjacent habitat and roadside management. Acta Oecologica **20**(1), 1–13 (1999)
41. Steen, D.A., et al.: Relative vulnerability of female turtles to road mortality. Anim. Conserv. **9**, 269–273 (2006)
42. Červinka, J., Riegert, J., Grill, S., Šálek, M.: Large-scale evaluation of carnivore road mortality: the effect of landscape and local scale characteristics. Mammal Res. **60**(3), 233–243 (2015). https://doi.org/10.1007/s13364-015-0226-0
43. Orłowski, G., Nowak, L.: Factors influencing mammal roadkills in the agricultural landscape of South-Western Poland. Pol. J. Ecol. **56**(2), 283–294 (2006)
44. Hill, J.E., et al.: Roads do not increase carrion use by a vertebrate scavenging community. Sci. Rep. **8**, 16331 (2018)
45. Mumme, R.L., et al.: Life and death in the fast lane: demographic consequences of road mortality in the Florida Scrub-Jay. Conserv. Biol. **14**(2), 501–512 (2000)
46. Novaes, N.L.M., et al.: On a collision course: the vulnerability of bats to Roadkills in Brazil. Mastozoología Neotropical **25**(1), 115–128 (2018)
47. D'Amico, M., et al.: Vertebrate road-kill patterns in Mediterranean habitats: who, when and where. Biol. Cons. **191**, 234–42 (2015)
48. Coffin, A.W.: From roadkill to road ecology: a review of the ecological effects of roads. J. Transp. Geogr. **15**, 396–406 (2007)
49. Laurance, W.F., et al.: Reducing the global environmental impacts of rapid infrastructure expansion. Curr. Biol. **25**, R259–R262 (2015)
50. McGregor, R.L., Bender, D.J., Fahrig, L.: Do small mammals avoid roads because of the traffic? J. Appl. Ecol. **45**, 117–123 (2008)
51. Rio-Maior, H., et al.: Designing the landscape of coexistence: Integrating risk avoidance, habitat selection and functional connectivity to inform large carnivore conservation. Biol. Cons. **235**, 178–188 (2019)
52. Kociolek, A.V., et al.: Effects of road networks on bird populations. Conserv. Biol. **25**(2), 241–249 (2011)
53. Mazerolle, J.M., Huot, M., Gravel, M.: Behavior of amphibians on the road in response to car traffic. Herpetologica **61**(4), 380–388 (2005)

54. Muñoz, P.T., Torres, F.P., Megías, A.G.: Effects of roads on insects: a review. Biodivers. Conserv. **24**(3), 659–682 (2014). https://doi.org/10.1007/s10531-014-0831-2
55. Haddad, N.M., et al.: Habitat fragmentation and its lasting impact on Earth's ecosystems. Sci. Adv. **1**(2), e1500052 (2015)
56. Ryser, R., et al.: The biggest losers: habitat isolation deconstructs complex food webs from top to bottom. Proc. R. Soc. B **286**, 20191177 (2019)
57. Riley, S.P.D., et al.: A southern California freeway is a physical and social barrier to gene flow in carnivores. Mol. Ecol. **15**, 1733–1741 (2006)
58. Dunee, J.A., Williams, R.J.: Cascading extinctions and community collapse in model food webs. Philos. Trans. R. Soc. B **364**, 1711–1723 (2009)
59. Corlett, R.T.: Climate change in the tropics: the end of the world as we know it? Biol. Cons. **151**, 22–25 (2012)
60. Littlefield, C.E., et al.: Connectivity for species on the move: supporting climate driven range shifts. Front. Ecol. Evol. **17**(5), 270–278 (2019)
61. Trombulak, S.C., Frissell, A.C.A.: Review of ecological effects of roads on terrestrial and aquatic communities. Conserv. Biol. **14**(1), 18–30 (2000)
62. Francis, C.D., Baber, J.D.: A framework for understanding noise impacts on wildlife: an urgent conservation priority. Front. Ecol. Environ. **11**(6), 305–313 (2013)
63. Mathevon, N., et al.: Singing in the rain forest: how a tropical bird song transfers information. PLoS ONE **3**(2), e1580 (2008)
64. Longcore, T., Rich, C.: Ecological light pollution. Front. Ecol. Environ. **2**, 191–198 (2004)
65. Shannon, G., et al.: Road traffic noise modifies behavior of a keystone species. Anim. Behav. **94**, 135–141 (2014)
66. Bhat, S.: Ecotoxicology and impact on biodiversity. J. Pharmacognosy Phytochem. **2**(2), 1–19
67. Laurance, W.F., Goosem, M., Laurance, S.G.W.: Impacts of roads and linear clearings on tropical forests. Trends Ecol. Evol.Evol. **24**(12), 659–669 (2009)
68. Laurence, W.F., et al.: A global strategy for road building. Nature **513**, 229–232 (2014)
69. Barber, C.P., et al.: Roads, deforestation, and the mitigating effect of protected areas in the Amazon. Biol. Cons. **177**, 203–209 (2014)
70. Wu, S., Li, B.V.: Sustainable linear infrastructure route planning model to balance conservation and socioeconomic development. Biol. Cons. **266**, 109449 (2022)
71. Fischer, T.B.: Strategic environmental assessment and transport planning: towards a generic framework for evaluating practice and developing guidance. Impact Assess. Project Appraisal **24**(3), 183–197 (2012)

Energy Efficiency Optimization and Carbon Emission Reduction

Comparative Study on Carbon Dioxide Heat Pump Design Under the "Double Carbon" Goal

Jun Zhang[1], Yali Wang[2](✉), Yahuan Lu[1], and Xin Xiong Liu[2]

[1] Shenzhen Gas Corporation Ltd., Shenzhen, China
[2] Huazhong University of Science and Technology, Wuhan, China
aliawang1001@163.com

Abstract. As China actively develops and implements environmental policies to achieve 'carbon peak' and 'carbon neutrality' to address global climate change, Carbon Dioxide heat pump technology has become crucial because it enhances energy efficiency, reduces greenhouse gas emissions, and integrates renewable energy resources. Based on 25 years of literature from the Web of Science and China National Knowledge Infrastructure databases, this study uses CiteSpace visualization tools and bibliometric and statistical methods to comprehensively analyze the international research dynamics on carbon dioxide heat pumps. Research abroad focuses on system optimization, design, and the enhancement of transcritical cycle performance, while China concentrates on transcritical cycle performance, compressor design, and regenerator performance. Additionally, international research trends shift from theoretical studies to technological applications, whereas research in China emphasizes the evolution of research paradigms and diversified research subjects. Significant differences exist between domestic and international research regarding key technologies, performance optimization, application areas, and methodologies. Overall, research in China has gradually aligned with international standards, particularly in empirical studies of technological performance in specific environments. Research should enhance international cooperation, explore system performance and cycle mechanisms more deeply, broaden research perspectives, and promote industry, academia, and research integration to better adapt to the deepening dual-carbon goals.

Keywords: Dual Carbon Goals · Carbon Dioxide Heat Pump · CiteSpace · Knowledge Graph

1 Introduction

The climate effect issue has always been one of the most concerning environmental issues in the world [1], and the content of carbon dioxide in the atmosphere is an essential factor affecting changes in atmospheric temperature. It has led people to reflect: After world-scale industrialization, will the rapid growth of carbon dioxide emissions caused by human activities affect the speed of global climate change? Will it intensify and trigger catastrophic climate effects? Therefore, the global community has adopted

G. Huang et al. (Eds.): ICESD 2024, ESE, pp. 93–110, 2025.
https://doi.org/10.1007/978-3-031-88683-6_8

measures to deal with climate change, including the United Nations Framework Convention on Climate Change, the Kyoto Protocol [2], the Bali Road Map, the Copenhagen Agreement and the Paris Agreement [3]. As an industrialized country with the largest population in the world and high carbon emissions, China regards climate effects as the biggest challenge in realizing socialist modernization and climate governance as the most significant opportunity for future social development. The Chinese government announced China's climate change policy in 2020, including achieving a peak in carbon emissions before 2030 and striving to achieve carbon neutrality by 2060 [4]. In the context of the "double carbon" goal, China has strengthened the control of carbon emissions, paying particular attention to energy consumption in the building sector, in which hot and cold water supply plays a vital role in building energy consumption [5]. Heat pump technology, especially the research on carbon dioxide heat pumps, has therefore received more and more attention.

Research on heat pump technology originated in 1912 and was initially used for house heating. In the 1970s, this technology began to be widely used in Europe and North America. For example, the annual output of heat pumps in the United States reached 500,000 units in 1977, and in 1998, air-source heat pump hot water units accounted for about 30% of hot water supply equipment [6]. China's research started relatively late, and it was not until 2001 that it began to pay attention to heat pump technology. In 2002, air-source heat pumps entered the southern market, mainly used in hotels, factories, and campuses. However, society's recognition of this technology was low at that time. 2008, the government issued relevant usage guidelines, marking policy support for the development of heat pump technology [7]. Of course, the progress of modern heat pump technology is inseparable from the development of refrigeration engineering. The development of refrigerants has gone through four stages [8, 9] and is currently developing towards environmentally friendly refrigerants such as CO_2 and H_2O. As a natural working fluid, CO_2 is easy to obtain, clean meets environmental protection requirements, has the characteristics of high density and low viscosity, and is suitable for efficient and stable heat pump systems [10, 11] (See Fig. 1).

Stage	First Stage	Second Stage	Third Stage	Fourth Stage
Date	1830-1930	1931-1990	1991-2010	2010-
Common Refrigerant	CO_2, SO_2, H_2O, Ethers, Ammonia, Air	Hydrochlorofluorocarbons (HCFCs), Chlorofluorocarbons (CFCs), H_2O	Hydrochlorofluorocarbons (HCFCs), Chlorofluorocarbons (CFCs), H_2O, CO_2	CO_2, Ammonia, Air, H_2O
Main Features	As long as it can be used	Security and durability	Protect the ozone layer	Low ODP, low GWP Prevent global warming Energy efficient

Fig. 1. Development Stages and Characteristics of Refrigerants

Research on carbon dioxide heat pumps has a long history worldwide. In contrast, China started late in this field and still needs sufficient systematic research, leaving a gap with developed countries. Therefore, this study uses CiteSpace [12] to conduct a knowledge graph analysis of relevant literature in Web of Science (WOS) and China

National Knowledge Infrastructure (CNKI), aiming to comprehensively examine the evolution and trends of carbon dioxide heat pump research at home and abroad. To promote global technological development, improve energy efficiency, promote sustainable energy utilization, and achieve sustainable development of the economy, environment and society.

2 Research Design

2.1 Methods

This study employs the visualization software CiteSpace, combined with bibliometric methods such as co-citation, clustering, and social network analysis, to create visual maps that illustrate the key pathways and frontier trends in carbon dioxide heat pumps. Given the limitations of CiteSpace in accurately identifying methodologies and theoretical foundations within the literature, content analysis was also applied to re-code and statistically analyze research themes, theoretical foundations, and methodologies. Additionally, quantitative data within the literature was examined in depth using statistical analysis methods and the SPSS tool, ensuring a comprehensive grasp of domestic and international carbon dioxide heat pump research.

2.2 Data Sources

International literature data was meticulously sourced from the Web of Science Core Collection. Advanced search queries using "TS = (Carbon Dioxide Heat Pump) OR TS = (CO_2 Heat Pump) OR TS = (Carbon Dioxide Heat Pump System)" were employed to filter publications between January 1988 and December 2022. This search was restricted to articles only, initially retrieving 3,399 records. To ensure data quality and reliability, these records were locked into the SCI (Science Citation Index) and duplicates were removed, culminating in a final tally of 2,599 papers. Domestically, the CNKI database was queried with keywords "carbon dioxide heat pump" or "CO_2 heat pump design" yielding 770 related documents. After filtering out book reviews, news reports, and conference notifications for the period from 1998 to 2022 and focusing on key Chinese academic indexes like the Peking University core, CSSCI, and EI, a total of 203 core domestic papers met all selection criteria. The selection of these papers was based on their relevance to the research topic, their quality, and their impact in the field.

2.3 Methodological Framework

This paper explores the knowledge structure and trends in carbon dioxide heat pump research using bibliometric and statistical analysis across four stages: Initially, data is prepared by collecting, deduplicating, categorizing, and cleaning literature from the WoS Core and CNKI databases, then imported into CiteSpace for network construction. Analytical phases utilize CiteSpace's LSI and LLR algorithms for longitudinal analysis of publication metrics and trends, while the Burstness model and SPSS software examine the evolution of research hotspots. Comparative analysis identifies similarities and differences in research approaches internationally and within China, enriching the understanding of each region's focus and strengths. Conclusions summarize the state of the research field, proposing directions aligned with China's dual-carbon goals to advance technology and support sustainable development. The methodological framework and stages are detailed in Fig. 2.

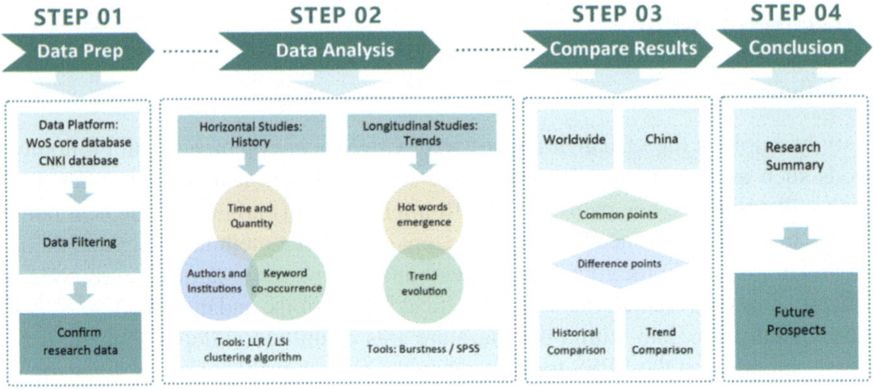

Fig. 2. The Research Framework and Process Design

3 Cross-Sectional Study of Carbon Dioxide Heat Pump Research: Historical Perspective

3.1 Research Timeline and Volume of Results

Over the past two decades, global research on carbon dioxide heat pumps has surged, with the Web of Science Core Collection documenting 2,599 SCI-indexed papers from 1998 to 2022. In 2022 alone, a record 385 papers were published, marking an all-time high. However, Chinese research began later, with only 1–2 papers annually between 1998 and 2007. A pivotal study by Yitai et al. [13] in 1998 highlighted the thermal recovery benefits of CO_2 transcritical cycles, spurring growth post-2007. By 2020, China's annual SCI paper output exceeded 100, boosted by national dual-carbon goals. In three years, publications soared to 416, positioning China as an emerging leader in this field (See Fig. 3).

Global research on carbon dioxide heat pumps is surging, led by China with over 700 publications, surpassing early starters like the US, Germany, and Japan. By 2022, China emerged as the leader, followed by the US with 300 SCI-indexed papers, and other developed nations. Thus, it is evident that China has firmly established itself as the leader in carbon dioxide heat pump research, as shown in Fig. 4.

3.2 Analysis of Authors and Institutions

Despite starting later, China now leads global carbon dioxide heat pump research, with Xi'an Jiaotong University at the forefront with 125 publications. Other leading institutions include Tianjin University, IITs, Swiss Federal Institute of Technology, and the US

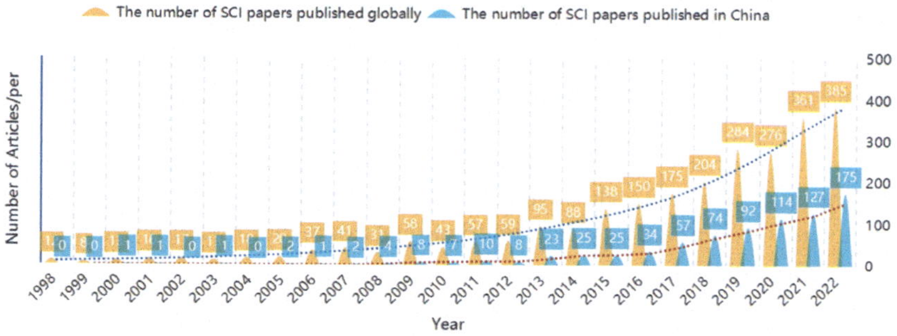

Fig. 3. Trend Chart of the Number of Publications Related to Carbon Dioxide Heat Pump-related Research at Home and Abroad from 1998 to 2022

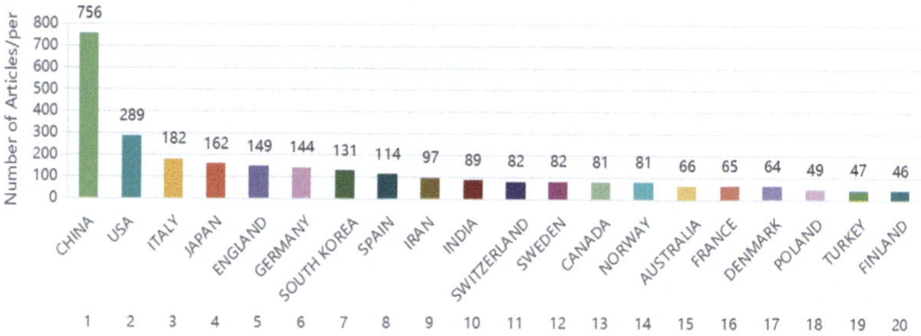

Fig. 4. Number of Publications on Carbon Dioxide Heat Pump Research in Various Countries Around the World from 1998 to 2022

Department of Energy. Academia drives this research, but the University of California excels in influence, with a single paper cited 285 times, indicating highly focused and collaborative international research, as shown in Fig. 5.

Fig. 5. Institutional Distribution Map of International Carbon Dioxide Heat Pump Research from 1998 to 2022

In China, the Institute of Thermal Energy at Tianjin University leads in publications with 19 papers, followed by critical universities like Shanghai University of Science and Technology and Xi'an Jiaotong University (Fig. 6). Domestic collaboration and diversity in research branches must meet international standards, with most studies concentrated in major universities' engineering departments. Notably, Ma Yitai of Tianjin University authored the most papers, 22. However, the Institute's papers have no citations, reflecting the need to enhance research quality, international cooperation, and researcher training to boost global visibility.

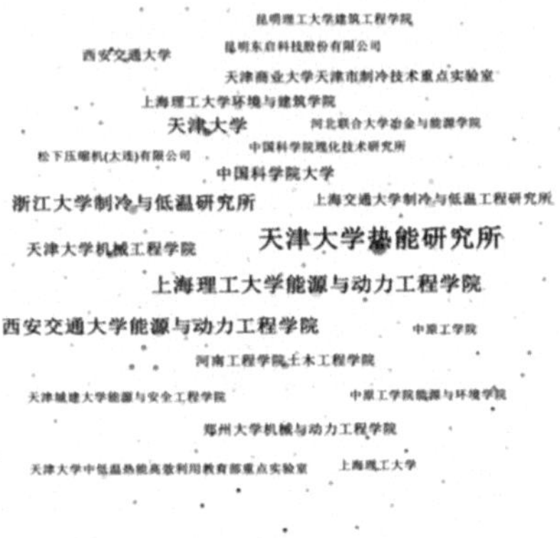

Fig. 6. Institutional Distribution Map of Carbon Dioxide Heat Pump Research in China from 1998 to 2022

3.3 Keyword Clustering

Keyword clustering in international carbon dioxide heat pump research

This research used the Latent Semantic Indexing (LSI) algorithm within CiteSpace software to systematically cluster keywords from the international literature on carbon dioxide heat pumps to delineate the main research directions. The analysis revealed a modularity score (Q) of 0.3894 and an average silhouette score (S) of 0.6781, which exceeded the predefined thresholds of 0.3 and 0.5, respectively. These results indicate that the obtained clustering structure is significant, and the clustering results are reasonable. Seven cluster labels were identified after filtering the results in Cluster Explorer, as shown in Table 1.

To present the clustering results visually, a keyword clustering visualization map was generated using CiteSpace, as shown in Fig. 7. These findings are significant as they reveal that the international research on carbon dioxide heat pumps includes seven main topics: carbon dioxide, Renewable energy, CO_2 capture, organic Rankine cycle, Energy efficiency, Extractive distillation, and Steady-state. This comprehensive categorization provides a clear overview of the current research landscape in this field.

Internationally, carbon dioxide heat pump research has long been a focal point in academia, focusing primarily on technical performance and system optimization, application scenarios, and empirical analysis. This research aims to improve energy efficiency,

Table 1. Cluster Analysis of International CO_2Heat Pump Research Literature

Cluster number	Silhouette value	Cluster name	High-frequency keywords
#0	0.632	Carbon dioxide	Optimization; System; Design; Refrigerants; Transcritical cycle
#1	0.574	Renewable energy	CO_2 mitigation; Chemical sector; Industrial decarbonization; Exergo-economic cost; District heating
#2	0.631	CO_2 capture	Energy storage; Power load; Steam mixture cycle; Recompression brayton cycle; Thermal energy storage
#3	0.689	Organic rankine cycle	Ground source heat pump; Sensitivity analysis; cchp system; Operational mode; Solar energy
#4	0.796	Energy efficiency	Geothermal energy; Energy service; Project finance; Sustainable farming; Heat recovery
#5	0.925	Extractive distillation	Heat integration; Energy saving; Ethyl acetate; Partial condensation; CO_2 emissions
#6	0.977	Steady state	Heat transports; Atmospheric CO_2; Dioxide; Gas exchange; Ocean acidification

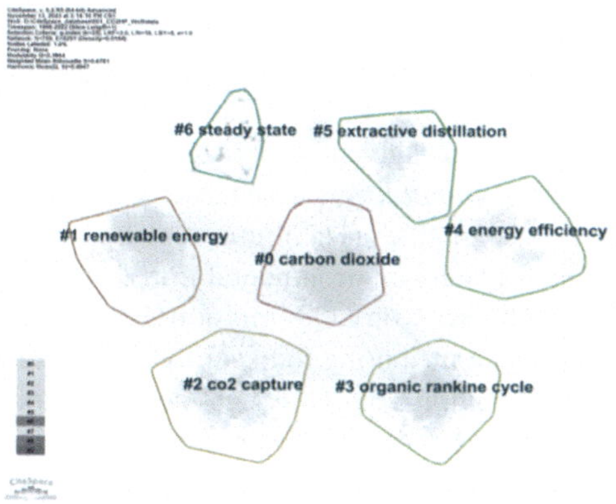

Fig. 7. Clustering visualization of international carbon dioxide heat pump research literature

optimize the system design, and deepen the understanding of the effects and feasibility of practical applications. Accordingly, based on the clustering results, this paper categorizes the international research on carbon dioxide heat pumps into four research dimensions: Optimization and System Design, Renewable Energy and Carbon Mitigation, CO_2 Capture and Storage, and Energy Efficiency and Sustainable Development, as shown in Fig. 8. The construction of this multidimensional research framework for carbon dioxide heat pumps helps researchers to systematically understand the current state of international research in this field. It provides a valuable tool for future research planning and collaboration.

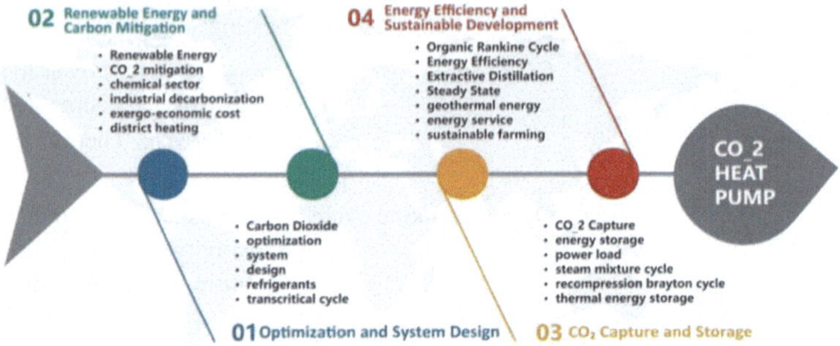

Fig. 8. Deconstruction Framework Diagram of International Multi-Dimensional Research on CO_2 Heat Pumps

Optimization and System Design: Emphasizes efficiency improvements through refrigerant selection and transcritical cycle optimization, incorporating aspects like heat exchanger design and system performance innovations [14].

Renewable Energy and Carbon Mitigation: Explores the integration of CO_2 heat pumps in renewable energy systems, focusing on their role in decarbonizing the chemical and industrial sectors, and the application of transcritical CO_2 pumps in novel hybrid energy systems for enhanced thermal sharing in low-temperature networks [15].

CO_2 Capture and Storage: Addresses the enhancement of CO_2 capture and storage efficiency through technologies like the Calcium Looping (CaL) process, which presents a promising solution for energy storage and CO_2 capture in power plants [16].

Energy Efficiency and Sustainable Development: Covers the use of geothermal energy, project financing, sustainable agriculture, and heat recovery to improve the energy efficiency, sustainability, and environmental impact of CO_2 heat pump systems. Studies indicate significant influences on CO_2 emissions from factors like national energy mix and heat pump efficiency, contributing to decarbonization strategies [17].

The diverse approaches to improving CO_2 heat pump technology, as illustrated in this paper, underscore the global commitment to advancing energy efficiency, renewable integration, and environmental sustainability. This framework provides insights into current trends and paves the way for potential innovations in the field, supporting researchers and policymakers in their endeavours.

Keyword Clustering in Domestic CO_2 Heat Pump Research

This study initially selected 286 keywords through systematic retrieval of domestic literature. After statistical analysis and eliminating or merging redundant keywords,

92 distinct keywords were identified. Further in-depth analysis of the frequency and relevance of these keywords led to their categorization into 21 groups, as shown in Fig. 9.

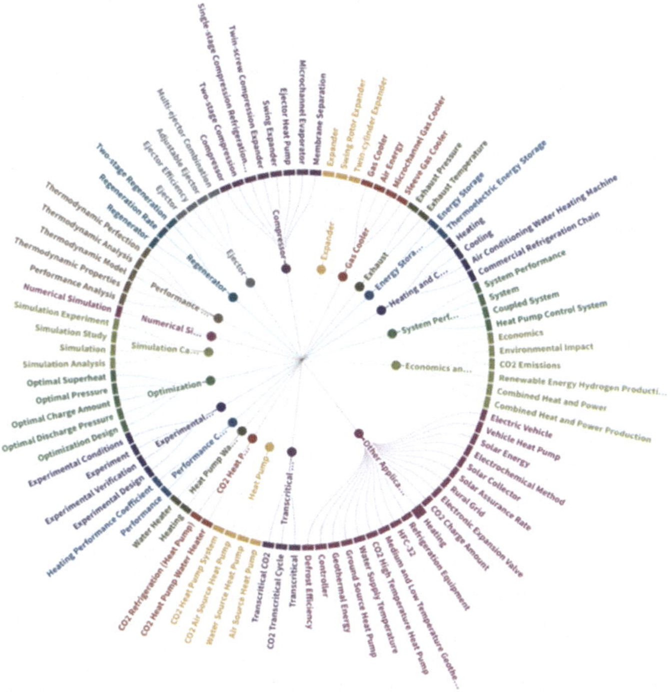

Fig. 9. Keyword Clustering Map of CO_2 Heat Pump Research in China

Compared to international research, the volume of core papers published domestically is relatively minor, but the research themes are more holistic. Based on the 21 groups of keyword clusters, this study categorizes the research on domestic CO_2 heat pumps into three primary research directions. It constructs a deconstruction logic diagram of domestic CO_2 heat pump research directions, shown in Fig. 10.

CO_2 Heat Pump Technology: The advancements in CO_2 heat pump technology, a critical focus in China, have been significant. Research primarily revolves around transcritical cycles and heat pump water heaters. Early studies, such as those by Junlan et al. [18], analyzed CO_2 transcritical refrigeration cycle systems and identified that reducing throttling losses could significantly boost system efficiency. Subsequent research by Hongtao et al. [19] on transcritical cycles in heat pump water heaters showed that CO_2-based systems could achieve a Coefficient of Performance (COP) of 4.3, with an average annual COP of 4.0, yielding energy savings up to 75% over conventional systems. This area involves novel cycle design and system integration to enhance performance and expand applications.

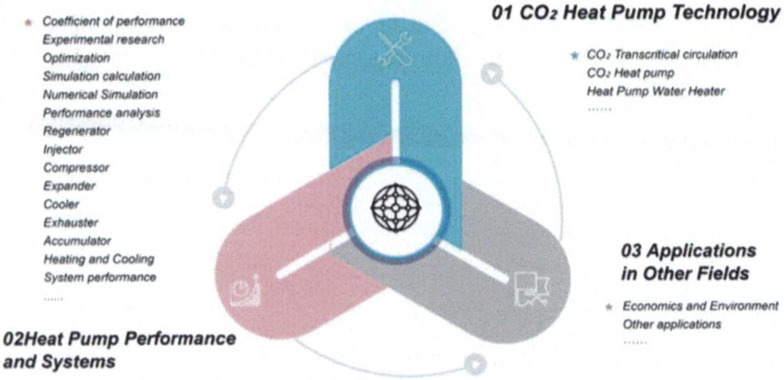

Fig. 10. Deconstruction Logic Diagram of China's CO_2 Heat Pump Research Direction

Heat Pump Performance and Systems: Efforts in this area are targeted at improving performance through coefficients of performance, experimental research, and optimization. Liu Changchun et al. [20] demonstrated the effective operation of CO_2 heat pump water heaters under varying seasonal conditions, achieving performance coefficients as high as 5.9. Li Minxia et al. [21] also emphasized the importance of optimizing operational parameters in their studies on CO_2 transcritical cycle water source heat pumps.

Applications in Other Fields: The potential of CO_2 heat pump technology is not limited to traditional areas. Researchers are also applying it in areas like electric vehicles and renewable energy, opening up new possibilities. Ming et al. [22] showed significant enhancements in both cooling and heating performance of intermediate cooling type CO_2 heat pump systems in electric vehicles. Moreover, Ting et al. [23] implemented a "data center+" energy utilization scheme that turns data centers from energy consumers into providers by using waste heat in regional heating and cooling systems, demonstrating substantial economic and energy efficiency benefits.

In summary, comprehensive research and detailed keyword analysis have outlined the primary directions in domestic CO_2 heat pump research: advancing technology, enhancing system performance, and broadening applications to fields like electric vehicles and solar energy. This structured approach aids in understanding the evolving landscape and driving further innovation in CO_2 heat pump technology.

4 Longitudinal Research on CO_2 Heat Pumps: Focus Areas and Evolution

Using the CiteSpace tool, we have honed a method to decipher CO_2 heat pump research trends. Its keyword highlight and timeline maps are crucial, particularly the former, which spotlight research frontiers and forecast future directions by analyzing time spans and intensity shifts. These maps are invaluable for navigating complex foreign literature and identifying patterns. On the other hand, timeline maps effectively reveal shifts in domestic research, enhancing our understanding of domestic and international trends. This approach leverages tool benefits to tackle the variation in research volume and complexity, ensuring thorough analysis.

4.1 Research Hotspots and Evolutionary Characteristics of Foreign CO_2 Heat Pumps

This study highlights the critical keyword distribution and primary research directions in domestic CO_2 heat pump technology through extensive literature retrieval and keyword analysis. Research areas are divided into three main categories: advancing CO_2 Heat Pump Technology, enhancing Heat Pump Performance and Systems, and exploring applications in other fields like electric vehicles and solar energy. This framework provides researchers with insights into the current landscape and supports the practical application of CO_2 heat pumps across various sectors.

Internationally, CO_2 heat pump research began earlier and featured more complex analyses. Using Citespace's Burstness model, 25 significant keyword groups were identified and then analyzed with SPSS for time series, frequency statistics, and correlation measurements. Further analysis using R's Bibliometrix package and Sankey diagrams elucidated the dynamics and trends in international CO_2 heat pump research (See Fig. 11).

Fig. 11. Sankey Diagram of Research Trends in International CO_2 heat Pump Field

Deepening from Theoretical Research to Applied Practice: Early research on carbon dioxide (CO_2) heat pumps primarily focused on fundamental concepts such as "Carbon Dioxide" and "Chemical Heat Pump," examining their relationship with emerging technologies like the "Transcritical Cycle" [24]. In this cycle, the working fluid, CO_2, exists in gas and liquid states above its critical point, enabling highly efficient heat exchange at elevated temperatures. As theoretical understanding improved, the focus expanded to application-specific areas like "Heat Exchangers" and "Performance Characteristics" [25], emphasizing the impact of heat exchanger design and performance on overall system efficiency under various operational conditions.

Progressing from Technology Prediction to Verification and Optimization: Initial CO_2 heat pump studies concentrated on technical aspects such as heat transfer, pressure drop, and experimental data. Subsequent research incorporated more sophisticated concepts like the "Thermodynamic Cycle" [27], marking a transition from predictive to verification stages. This evolution broadened the research to include comprehensive technical elements like mass flow rate and Combined Cooling, Heating and Power (CCHP) systems [28], focusing on optimizing system performance and integrating multifactorial performance evaluations.

Expanding Comprehensively from One-dimensional to Multidimensional Analysis: Initially, studies on CO_2 heat pumps targeted basic cyclic processes conducted on one-dimensional planes. Introducing concepts like the "Internal Heat Exchanger" [29] transitioned research into a two-dimensional framework. Later, the "Dividing Wall Column" [30] concept shifted towards more holistic system-level considerations, extending the scope of research into three-dimensional analysis. This expansion highlights the adaptation of physical heat exchange processes and virtual processes enabled by technological innovation, significantly advancing the research landscape of CO_2 heat pumps.

These advancements underscore a significant shift in CO_2 heat pump research from basic theoretical frameworks to multidimensional and practical applications. This shift is not just a change in research focus, but a reflection of the increasing relevance and impact of CO_2 heat pump technologies in real-world settings. It's a testament to the field's deepening, from exploring fundamental concepts to addressing real-world applications and systemic optimizations, and it's a clear indication of the field's trajectory towards enhancing energy efficiency and sustainability in heat pump technologies.

4.2 Research Hot Spots and Evolution Characteristics of Domestic Carbon Dioxide Heat Pumps

Compared with foreign countries, domestic research on carbon dioxide heat pumps started late, and the research time limit is relatively short. As a result, 20 groups of domestic emergent keywords were obtained, and the age segmentation of the Sankey diagram was only divided into two stages. However, within this limited research time frame, the research literature on carbon dioxide heat pumps still shows specific trends (See Fig. 12).

Research paradigm evolution: In the early days of domestic carbon dioxide heat pump research in China, the focus was mainly on concepts such as "transcritical cycle" and "heat pump", focusing on technical verification and performance discussion. However, as time goes by, the research paradigm gradually evolves, and newly introduced keywords such as "expander" and "carbon dioxide transcritical cycle" indicate that the research is gradually deepening and involves more complex directions such as system performance and cycle mechanisms. For example, in 2021, the research of scholars such as He Yujia of Tongji University [31] simplified the control of the carbon dioxide heat pump water heater system through the fixed point optimization method of the charging amount, maintaining high energy efficiency operation while reducing the energy efficiency attenuation of the system. This change reflects the increasingly professional and in-depth research on carbon dioxide heat pumps.

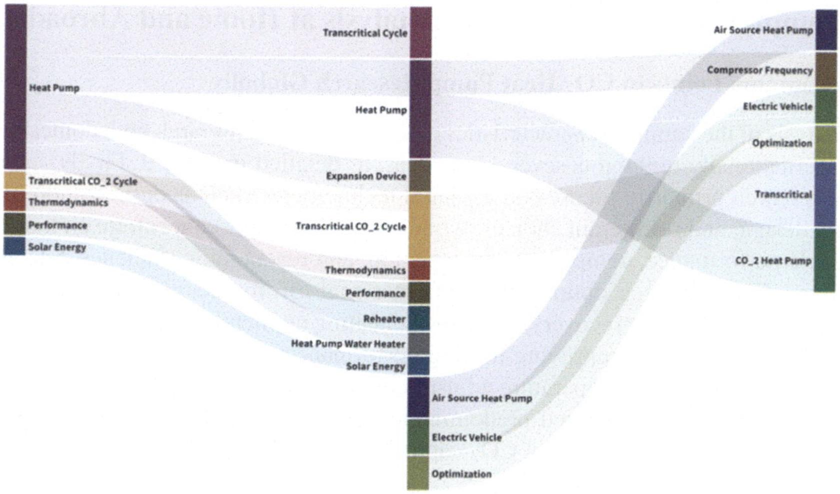

Fig. 12. Sankey Diagram of Research Trends in China's Field of CO_2 Heat Pumps

Changes in research subjects: The initial research subjects focused on conceptual nouns and gradually developed towards empirical research and application orientation. The emergence of keywords such as "regenerator" and "heat pump water heater" indicates that the research body is gradually focusing on the practical application and effect evaluation of carbon dioxide heat pumps in specific scenarios, showing a research trend of mainly empirical research. For example, scholars such as Xuan Fuchen and others [32] summarized the research progress of transcritical CO_2 refrigeration systems and proposed methods to improve refrigeration efficiency and heat transfer efficiency. They discussed its application in automobile air conditioners, heat pumps, food refrigeration and refrigeration—future research directions. The evolution of this research body shows that researchers pay more attention to practical application value and effect evaluation.

Expansion of research scenarios: From the perspective of research scenarios, the initial discussion of system concepts gradually evolved into a focus on specific scenarios and application environments. The introduction of keywords such as "solar energy" and "air source heat pump" reflects the gradual expansion of research scenarios into more comprehensive and complex environments, showing researchers' interest in carbon dioxide heat pumps in diverse scenarios. For example, scholars such as Wanyong [33] emphasized the criticality of optimizing the losses of the condenser and compressor in the system in complex application scenarios in the research results of the performance test of the electric vehicle's heat pump air conditioning system. This trend shows that future research may focus more on practical engineering applications and in-depth study of different application scenarios.

5 Comprehensive Comparative Analysis at Home and Abroad

5.1 Common Points in CO_2 Heat Pump Research Globally

An analysis of the common characteristics of CO_2 heat pump research both domestically and internationally highlights several key areas, as detailed in Table 2. Firstly, international research primarily focuses on essential technologies such as optimization plans, system design, and refrigerant choices, while domestic studies concentrate on transcritical cycle performance, compressor optimization, and regenerator efficiency. Secondly, in terms of performance optimization, international efforts are geared towards CO_2 emission reduction, heat recovery, and district heating applications. In contrast, domestic research emphasizes improving heating performance coefficients, exhaust pressure optimization, and water temperature adjustments. Thirdly, regarding application fields, international research is applied predominantly in the chemical sector and renewable energy, with a strong emphasis on CO_2 capture and storage.

Table 2. Common Characteristics of CO_2 Heat Pump Research at Home and Abroad

Common characteristics	International research	Domestic research
Key technologies	Optimization; System Design; Refrigerants; Transcritical Cycle	Transcritical cycle performance; Compressor design and optimization; Regenerator performance research
Performance optimization	CO_2 Mitigation; Heat Recovery Optimization; District Heating	Enhancement of heating performance coefficient; Optimal exhaust pressure design; Supply water temperature regulation
Application areas	Chemical Sector; Renewable Energy Applications; CO_2 Capture and Storage	Air source heat pump applications; Water source heat pump technology; CO_2 heat pump water heater applications
Research methods	Cluster Analysis; Sensitivity Analysi; Operational Mode Analysis	Experimental verification of key parameters; Numerical simulation for performance prediction; Simulation and numerical modeling studies
Environmental protection	Sustainable Farming; Heat Recovery; CO_2 Emissions Reduction	Application of environmentally friendly CO_2 refrigerants

In contrast, domestic applications focus on practical uses such as air and water source heat pumps and CO_2 heat pump water heaters. Moreover, methodological diversity is noted; international studies employ cluster and sensitivity analysis for a multi-angled approach, while domestic research focuses on experimental verification and numerical simulations for performance predictions. Lastly, regarding environmental protection, international research highlights sustainable practices like farming and heat recovery for CO_2 emission reductions, whereas domestic work prioritizes using environmentally friendly CO_2 refrigerants. These commonalities outline the global concerns in CO_2 heat pump research across technology, performance, and application, offering valuable insights for future endeavours.

5.2 Differences Between China and International Research

Comparative analysis of carbon dioxide heat pump research reveals distinct trends between domestic and international efforts (see Table 3). Internationally, research began earlier and has seen exponential growth over a prolonged period. Domestically, despite a later start, research rapidly advanced beyond initial technical verification into performance research and empirical analysis, achieving significant growth in 2020. Internationally, major research entities include universities, scientific institutions, and energy departments, focusing on optimization, system design, renewable energy, carbon mitigation, CO_2 capture, and sustainable development. Domestically, the research landscape is more varied, encompassing higher education, university institutes, and private enterprises, reflecting a blend of industry, academia, and research. Domestic keyword clustering predominantly concentrates on CO_2 heat pump technology, system performance, and diverse applications. Regarding trends, international research transitions from theoretical studies to practical applications, moving from technology forecasts to multidimensional optimization. Conversely, domestic research evolves across paradigms, participant profiles, and scenarios, indicating a broader and more diverse trajectory in research development.

By comparing the above dimensions, there are significant differences in the start-up time, research institutions, keyword clustering and carbon dioxide heat pump research trends at home and abroad. This difference reflects the different orientations of various countries regarding research priorities, development stages and cooperation models in this field. Overall, domestic carbon dioxide heat pump research has successfully caught up with the pace of international development through rapid paradigm evolution, expansion of research perspectives, diversification of subjects, and wide application of scenarios. In the future, domestic research is expected to deepen exchanges and cooperation with international research further and promote continuous innovation and development in this field.

6 Conclusions

International vs. China Focus: The literature focuses on optimizing system design and transcritical cycle performance internationally. Domestically, the focus is shifting to transcritical cycle performance, compressor design and optimization, and regenerator performance. Research Trends: Internationally, the literature trend moves from theoretical research to practical application, from technology forecasting to optimization verification, and from one-dimensional to multi-dimensional developments. Domestically, there is a focus on the evolution of research paradigms, changes in research units, and the

Table 3. Comparison of Multi-Dimensional Differences in CO_2 Heat Pump Research at Home and Abroad

Dimension	International research	Domestic research
Time and Volume	Started earlier; Longer duration; Exponential growth in publications	Started later; Shorter duration; Significant increase in publications in 2020
Authors and Institutions	Dominated by higher education institutions; Research organizations; National energy departments	Higher education institutions; University research institutes; Private enterprises
Keyword Clustering	Optimization and System Design; Renewable Energy and Carbon Mitigation; CO_2 Capture and Storage; Energy Efficiency and Sustainable Development	CO_2 Heat Pump Technology; Heat Pump Performance and Systems; Applications in Other Fields
Research Trends	From theoretical research to practical application; From technological forecasting to optimization verification; From one-dimensional to multi-dimensional	Evolution of research paradigms; Changes in research entities; Expansion of research scenarios

expansion of research scenarios, reflecting a diversified trend. Differences in critical technologies, performance optimization, and application areas: There are notable differences between domestic and international research regarding key technologies, performance optimization, application areas, and research methods. Overall, international research on CO_2 heat pumps mainly focuses on technical optimization, system design, and transcritical cycle performance, which shows a more systematic and in-depth approach. Domestic research builds on technical studies through rapid paradigm shifts, expanded research perspectives, diversification of research units, and broad application scenarios, successfully catching up with the international pace. On the one hand, domestic research emphasizes performance optimization and application-oriented empirical research. On the other hand, it also shows unique strengths in key technologies and application areas. These differences reflect each region's unique focus and development path in CO_2 heat pump research and provide valuable insights for future in-depth studies and international collaborations.

Future Prospects: As the dual carbon goals deepen, further international collaboration and exchange in CO_2 heat pump research are expected. This will provide broader opportunities for technological innovation, policy support, and international cooperation. Your role in this future is crucial. Future research should focus more on the systemic performance and cycle mechanisms of CO_2 heat pumps. Moreover, research perspectives should be broadened to include both conceptual distinctions and a swift transition to empirical and application-oriented studies. Research entities should further diversify, with domestic studies potentially expanding into the business sector to foster closer industry-academia-research collaborations, thus better serving practical applications. In summary, CO_2 heat pump research is poised to take firmer steps in the pace of international development, offering more innovative and sustainable solutions to global climate change challenges.

References

1. Mingxing, W., Qingcun, Z.: Carbon dioxide content in the atmosphere. Atmos. Sci. **02**, 212–219 (1986)
2. Kyoto Protocol[EB/OL].(1997–12–11)https://unfccc.int/zh/kyoto_protocol
3. Paris Agreement[EB/OL].(2015–12–12)https://unfccc.int/process-and-meetings/the-paris-agreement
4. Jinping, X.: Delivered an important speech at the general debate of the 75th United Nations General Assembly. People's Daily 2020–09–23(001)
5. Marszal-Pomianowska, A., Zhang, C., Pomianowskim, M., et al.: Simple methodology to estimate the mean hourly and the daily profiles of domestic hot water demand from hourly total heating readings. Energy Build. **184**, 53–64 (2019)
6. Mengmeng, B., Lingyan, Y., Xinyu, Z.: Brief analysis of the development of ground source heat pumps in Europe. Constr. Sci. Technol. **02**, 73–78 (2019)
7. Shubiao, Y., Nianqing, Z.: Analysis of the development history of ground source heat pumps in China. Shanghai Land Resour. **38**(03), 57–61 (2017)
8. Jieshi, K., Xunchang, W.: The next generation of refrigerants—historical review, thinking and prospects. HVAC **39**(02), 41–49 (2009)
9. Wanjun, S., Chunli, L., Jianjun, Z., et al.: Research status and trends of mixed refrigerant standards. Refrig. Air Conditioning **20**(05), 6–9 2020
10. Biao, M., Zhaoliang, J.: Current research status and development prospects of carbon dioxide refrigerants. Refrigeration **31**(03), 36–43 (2012)
11. Linfeng, L., Ming, W., Ruijun, M., et al.: High-efficiency, energy-saving and environmentally friendly transcritical CO_2 heat pump technology. Sci. Technol. Innov. (03), 80–81+83 (2020)
12. Chen, C., Song, M.: Visualizing a field of research: A methodology of systematic scientometric reviews. PLoS ONE **14**(10), e0223994 (2019)
13. Yitai, M., Zhao, Y., Canren, L.: Thermodynamic analysis of CO_2 transcritical (reverse) cycle. J. Eng. Thermophys. (06), 665–668 (1998)
14. Yin, J.M., Bullard, C.W., Hrnjak, P.S.: Elements for the design of new generation of heat exchangers for transcritical CO_2 heat pumps. Air Conditioning Refrig. Center CR-37 (2001)
15. De Santoli, L., Lo Basso, G., Astiaso Garcia, D., et al.: Dynamic simulation model of trans-critical carbon dioxide heat pump application for boosting low temperature distribution networks in dwellings. Energies **12**(3), 484 (2019)
16. Ortiz, C., Valverde, J.M., Chacartegui, R., et al.: Scaling-up the calcium-looping process for CO_2 capture and energy storage. Kona Powder Part. J. **38**, 189–208 (2021)
17. Jadwiszczak, P., Jurasz, J., Kaźmierczak, B., et al.: Factors shaping A/W heat pumps CO_2 emissions—evidence from Poland. Energies **14**(6), 1576 (2021)
18. Junlan, Y., Yitai, M., Minxia, L., et al.: Performance evaluation and analysis of CO_2 transcritical refrigeration cycle system. Fluid Mach. **03**, 34–38 (2004)
19. Hongtao, X., Xiuling, Y., Guoqiang, L., et al.: Research on the application of transcritical circulating carbon dioxide in heat pump water heaters. J. Refrig. **03**, 12–16 (2001)
20. Shengchun, L., Yitai, M., Qiuju, L.: Experimental study on CO_2 heat pump water heater. J. Tianjin Univ. **02**, 238–242 (2008)

21. Minxia, L., Yitai, M., Lixin, L., et al.: Experimental study on operating characteristics of carbon dioxide transcritical cycle water source heat pump system with expander. J. Solar Energy **03**, 49–54 (2005)
22. Ming, X., Yiyu, C., Qinghui, Z., et al.: Experimental research on intercooling carbon dioxide heat pump system for electric vehicles. Automot. Technol. **02**, 43–47 (2022)
23. Ting, Z., Xiaodan, L.: Promote the transformation and application of carbon dioxide heat pump technology and promote the transformation of data centers from energy consumers to providers. Energy Saving Environ. Prot. **03**, 56–59 (2023)
24. Sarkar, J., Bhattacharyya, S., Gopal, M.R.: Transcritical carbon dioxide based heat pumps: Process heat applications (2004)
25. Carlson, M., Alvarez, F.: Design of a 1 MWth supercritical carbon dioxide primary heat exchanger test system. In: Energy Sustainability, vol. 83631, pp. V001T02A014. American Society of Mechanical Engineers (2020)
26. Massarotti, N., Mauro, A., Normino, G., et al.: Innovative solutions to use ground-coupled heat pumps in historical buildings: a test case in the city of Napoli, Southern Italy. Energies **14**(2), 296 (2021)
27. Dang, C., Hihara, E.: In-tube cooling heat transfer of supercritical carbon dioxide. Part 1. Experimental measurement. Int. J. Refrig. **27**(7), 736–747 (2004)
28. Ebrahimi, M., Majidi, S.: Exergy-energy-environ evaluation of combined cooling heating and power system based on a double stage compression regenerative gas turbine in large scales. Energy Convers. Manag. **150**, 122–133 (2017)
29. Zhu, Y., Li, C., Zhang, F., et al.: Comprehensive experimental study on a transcritical CO_2 ejector-expansion refrigeration system. Energy Convers. Manag. **151**, 98–106 (2017)
30. Kiss, A.A., Smith, R.: Rethinking energy use in distillation processes for a more sustainable chemical industry. Energy **203**, 117788 (2020)
31. Yujia, H., Xingyu, L., Jiahao, C., et al.: Fixed point optimization of carbon dioxide heat pump water heater charging volume. J. Eng. Thermophys. **42**(04), 824–827 (2021)
32. Fuchen, X., Jing, X.: Research progress on transcritical CO_2 refrigeration cycle system and application. Food Mach. **35**(08), 226–231 (2019)
33. Wanyong, L., Yusheng, L., Junye, S., et al.: Low-temperature heating performance and optimization of electric vehicle heat pump air conditioning system. J. Refrig. **41**(06), 31–39 (2020)

Optimized Human-Centric Decision-Maker for HVAC Systems: Using Genetic Algorithms to Establish an Equilibrium Point for Enhanced Thermal Comfort

Mahmoud I. Hozayen(✉), Wessam H. Abass, and Mohamed A. Fikry

Department of Architectural Engineering, Alexandria University, Alexandria, Egypt
{mahmoudhozayenh,wessam.abbas,mfikry}@alexu.edu.eg

Abstract. In the context of sustainable development, energy optimization has become an essential field of investigation, due to its vital role in the preservation of environment and energy efficiency. Significantly, in the domain of building services and operation, HVAC (Heating, Ventilation, and Air Conditioning) systems are among the most energy-consuming elements. Examining the root causes for discomfort and addressing them by refining operation of HVAC systems, this paper offers a decision-making framework to alter HVAC performance to be more-human centric and energy efficient, driving the focus on how occupant's comfort and building controls interact together. This is done by investigating a deficiency in current practices regarding the cooling of open layout plans. Current methodologies of sorting spaces into thermal zones often neglect the variations and localized impact of factors such as shading from adjacent buildings, variating direct sunlight values in space, and other factors. Subsequently, minor fluctuations in temperature are often overlooked in thermal zoning. As a result, conventional HVAC systems prove insufficient in capturing the nuanced dynamics of local temperature variations, which leads to inaccurate monitoring and control. By applying the concept of multi-agent control system to HVAC controls, this study argues that each space has an Equilibrium Point (EP) which achieves optimum thermal conditions and approaches the cooling strategy in a human-centric approach. Moreover, this EP is floating to adapt to the nuances and minor fluctuations. The research explores the effectiveness and feasibility of EP controls through simulation. Integrating a Computational Fluid Dynamics (CFD) model, automated and optimized using Genetic Algorithms (GA), to search for optimum controls for space within given context simulating the floating EP to achieve sustainable operation of HVAC. This decision-making tool helped HVAC systems be more efficient and optimized in cooling open layout spaces with 60% less fluctuations in temperatures than ordinary configurations. This research directly contributes to achieving Sustainable Development Goals (SDGs) by enhancing energy efficiency and reducing environmental impact.

Keywords: Thermal Comfort · Genetic Algorithms · Building Energy Modelling · Sustainability

G. Huang et al. (Eds.): ICESD 2024, ESE, pp. 111–122, 2025.
https://doi.org/10.1007/978-3-031-88683-6_9

1 Introduction

Thermal comfort is vital objective in building environments, critically influencing human health, productivity, and well-being [1]. As a result, HVAC systems are an inevitable element of contemporary buildings to supplement passive techniques [1]. But they consume substantial financial and energy expenditures. For example, on average, HVAC account for 46% of total energy use in USA [2]. This necessitates the optimization of HVAC systems not only to enhance cost-efficiency and occupant comfort but also to minimize environmental impact. By improving HVAC efficiency, we address a critical sector where potential gains in energy conservation and sustainability are both significant and essential [3]. This aligns with the goals outlined in The Sustainable Development Goals Report 2023, which emphasizes energy efficiency and environmental sustainability as key targets for global development.

It is argued that conventional HVAC practices often overlook the variance in heating and cooling needs around different spaces within a building [4, 5]. While thermal zoning is a known practice standards and codes still fall short to systematically provide standard methods of sorting thermal spaces. Nevertheless, even when implemented in efficient ways, the concept of thermal zoning fails to sufficiently account for the minor fluctuations that result from the dynamic surrounding environment [6].

While BMS protocols possess the ability to predict variations in occupancy and weather conditions, and adjust accordingly [7], they frequently overlook localized fluctuations within the space, which can result in areas being overheated or overcooled. This emphasizes the significance of developing an adaptive approach to HVAC management. One capable of adapting and evolving in response to the fluctuating diverse climatic demands of contemporary architectural environments. Therefore, the major aim of this research is to develop an innovative method that can effectively capture the intricacies and local variations that are inevitable within architectural environments, contributing to sustainable development by optimizing energy use and enhancing occupant comfort.

2 Literature Review

Building Energy Optimization has been approached diversely in literature. Depending on the discipline of the researcher background, their approach can fork usually into two: Prediction, and operation of HVAC, and Geometrical and passive controls. Firstly, predicting the thermal load of the HVAC systems contributed by large to reduction in energy use. Much research sought systematic ways to predict accurately the cooling and heating loads of spaces [8]. Numerous studies and research papers also describe sophisticated applications of Building Energy Management Systems (BEMS) [9, 10]. By using BMS, the Model Predictive Control framework (MPC) can accurately predict the thermal pattern and requirements of the building, with the use of computational models for forecasting, predicting, and extrapolating future scenarios [11].

Nevertheless, the use of advanced computational tools improved the operation of the HVAC systems. Such as lowering peak-to-average power demand ratio [12], improving the utilization of power assets [13], or enhancing the reliability of power systems using Demand Response method. On another hand, enhancing the operation or the control of the HVAC systems proved very efficient [14]. With the increasing popularity of Metaheuristic and search algorithms, some research offers optimal control algorithms [15]. Use of such algorithms in the improvement of control systems enhance energy management by reacting to unforeseen scenarios or forecasting system behavior [16].

Other researchers, however, approached the problem in an architectural manner: changing geometrical or semantic features of building for a better thermal behavior. It is widely argued that building energy use can be significantly cut off by employing passive means [17]. In this study, they parametrized building elements in a simulation framework using genetic algorithms to find the optimum thermal behavior. Changing the balcony type, dimensions of openings, glazing, system, etc. In [18], multi-objective optimization algorithms were used to extract optimal solutions targeting the conceptual design stage, which were validated crucial to indoor environment improvement and energy saving. Also, [19] investigated the impact of building parameters on energy consumption and determine the ideal design.

However, it is debated in [6] that one of the key concepts that is crucial in determining the HVAC controls is the thermal zoning. According to ASHRAE Standard 90.1–2016 [20], thermal zoning is defined as "a space or group of spaces within a building with heating and cooling requirements that are sufficiently similar so that desired conditions (e.g., temperature) can be maintained throughout using a single sensor (e.g., thermostat or temperature sensor).". Nevertheless, [6] raised the issue that there is a lack of studies and guidelines for methods of thermal zoning. Furthermore, [21] demonstrated some root causes that underscore the importance of thermal zoning, emphasizing the thermal variations inside spaces locally. [6] also showcased multiple studies that tried to fill in this gap and develop methods for capturing thermal fluctuations and adapting to them. However, most studies approach the thermal zoning as blocks of the layout plan with similar thermal loads, still overlooking the minor fluctuations in the same zones. Despite that, [22] have introduced a framework that utilizes the VAV technology to cool the space optimally considering each VAV terminal's effect to the overall thermal behavior, showing promising results for the framework.

3 Methodology

3.1 Equilibrium Point

Inspired by the Nash Equilibrium Point from the Game Theory by the Noble Laureate John Nash [23], this study suggests that each space possesses a unique, instantaneous Equilibrium Point (EP). A point where the controls achieve an optimal state, where any alterations to controls will yield worse thermal conditions. However, this EP characterizes by being 'floating', having a dynamic nature that changes continuously to counter the fluctuating conditions of space. Previous studies about VAV optimization using GA also approached the optimization of VAV controls of each space individually, though the dynamic capabilities of the system [24].

The implementation of EP ensures that the environmental controls evolve to maintain this state of equilibrium to the space, constantly shifting to cope with the requirements of its occupants. By utilizing VAV technology to approach the space as a single system, the

HVAC system will be treated as a multi-agent control system, where each VAV terminal affects the overall thermal behavior of the space.

To evaluate the EP concept, our methodology integrates optimization techniques (Genetic Algorithms), computer modelling, and simulations. Considering the complex and dynamic nature of built environments, this study focuses on its application within an open layout office plan. Conventional HVAC systems typically employ diffusers that distribute cooled air uniformly across a space. This approach, based on fixed temperature thermostats and setpoints, is where inefficiencies lay, failing to account for the diverse and dynamic thermal needs of different zones within the space [22].

The office will be virtually divided into a virtual 2D mesh. The goal is to maintain uniformity of temperature over all pixels. However, each pixel has its own cooling requirements, which are a function of its cardinal position. Therefore, we will implement the VAV technology—which involves the delivering chilled air through a primary duct system equipped with diffusers that regulate the airflow through each terminal. Physically, each terminal not only affects the pixels in proximity, but rather-according to Fick's Law of Diffusion and the turbulent nature of air flow-impacts the overall stability of the system.

As a result, the system will be treated as a multi-agent control system, designed to optimize the distribution of cooling considering all terminals altogether, rather than independently. This process of determining the optimal mass flow settings instantaneously presets a difficult computational challenge. Where the temperature at each pixel is considered a weighted function of all mass flow inlets. To accomplish this, the study utilizes Computational Fluid Dynamics (CFD) simulations carried out in ANSYS Fluent, which uses the finite volume method to model and assess the thermal behavior of the space under various air flow situations.

3.2 Optimization Approach

In light of the system's dynamic challenges, the problem was conceived as a search problem, driven by trial-and-error simulations. However, the conventional solution of trying every possible permutation would be unfeasible and impractical due to the enormous number of possible permutations of mass flow inlet values, which would require years of continuous computation to counter this, Genetic Algorithms (GA) was introduced to navigate the vast search space, since it is well suited for solving complex optimization problems.

Additionally, the study utilizes "Pyfluent", an API that enables programmatic interface with ANSYS Fluent, to automate the simulation environment dynamically. It facilitates adjusting variables and automating the iterative process of trial and error required by GA. This integration leverages the search and identification of the optimal configuration of HVAC controls, achieving EP. It also offers a robust, data-driven, and human centric thermal optimization altogether.

Performance Index. To evaluate the solutions: the baseline solution, and the proposed optimized solution, two methods were considered, numerical and visual.

The numerical method: A specific metric was developed: 'Performance Index (PI)' to provide a clear numeric value that reflects the uniformity of temperature distribution across the office space The PI is the standard deviation of each pixel's temperature from a predetermined target temperature, as in Eq. 1. A hypothetical surface at 1 m altitude above ground was designated to be the temperature harvesting surface.

$$PI = \sqrt{\frac{1}{n}\sum\nolimits_{i=1}^{n}(T_i - T_{target})^2} \tag{1}$$

This formula is calculated using the pixel's center temperature (T_i), target reference temperature (T_{target}), number of pixels (n). A decreased PI shows that most mesh pixels' temperatures are closely clustered around the target setpoint, indicating better thermal uniformity and precision. Thus, a lower Performance Index score suggests a better optimized system where each zone is at equilibrium, and vice versa. This metric quantifies system performance and permits careful simulation configuration comparison. This ensures that the evaluation of potential solutions is standardized in the search for optimal thermal comfort with minimal deviation from set thermal conditions. In addition to ensuring the precision and adaptability of the proposed solutions in achieving enhanced environmental control within dynamic architectural spaces.

The visual method: using the heat map, depicting temperature variations in plan using color gradients. This will enable recognition of the thermal distribution pattern throughout the plan, easily identifying areas of thermal inefficiency—such as hot and cold spots. Comparing heat maps from the baseline and optimized solution provides insight on how the Equilibrium Point (EP) model affects temperature distribution, hence HVAC system efficiency.

4 Simulation Environment

This section of the study outlines the simulation environment designed to mimic a realistic open layout office space. The simulated environment consists of a 15 m by 15 m office, with Variable Air Volume (VAV) terminals distributed evenly Fig. 1.

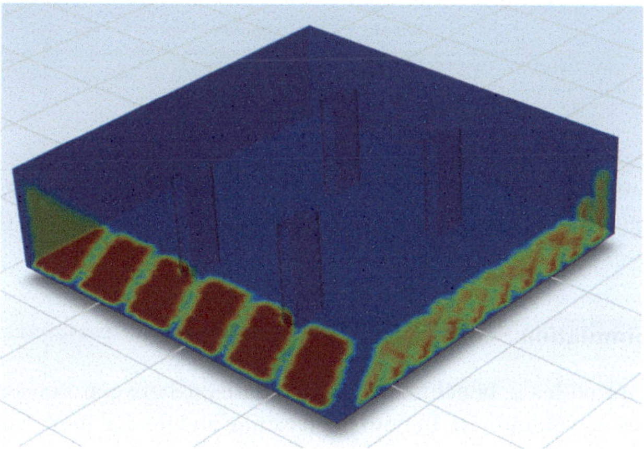

Fig. 1. Solar load Flux on Room, Perspective view (Ansys Fluent)

The office contains wide glazing on the western and southern facades, allowing for direct sunlight penetration, as well as with interior columns causing natural non-uniform thermal load in space. While large portion of literature depend on simulation tools such as "E-Plus" for building energy modelling, this study will use Ansys Fluent for its detailed computational fluid analysis capabilities [25, 26]. Using 'ANSYS Fluent 2022R2' as the simulation environment, the modelling, meshing, and solving were done by the built-in Fluent tools: 'DesignModeler', 'Fluent Meshing', 'Fluent Solver', respectively.

The meshing, by 'Fluent Meshing', provides the necessary granularity to accurately simulate the thermal dynamics within the office space. This step is crucial for ensuring that the computational models capture the intricate details of airflow and temperature distribution. The solving, by 'Fluent solver', applies physics models to accurately mimic real-world conditions. These include: *Energy Equations*: Applying Navier-Stokes equations to model the flow of heat and air within the office, *Turbulence Models*: Applying the k-epsilon model, suitable for turbulent flow prediction and thermal prediction accuracy, *Sun Ray Tracing*: Applying the surface-to-surface (S2S) model to account for the effects of direct sunlight entering through the windows, which significantly impacts the thermal comfort levels across different zones of the office Fig. 2.

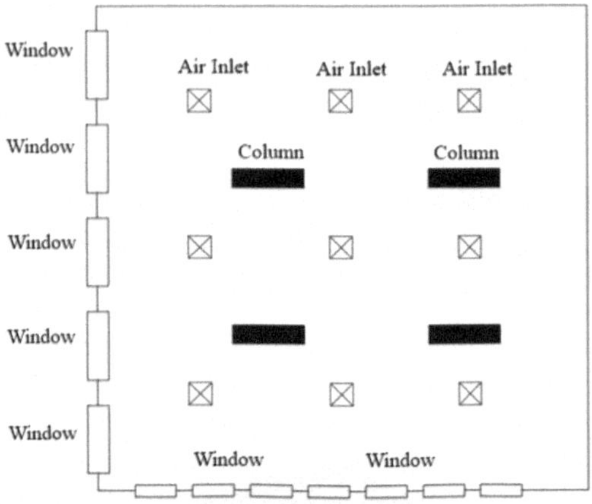

Fig. 2. Layout Plan

The simulation captures the sun angle and intensity to mimic the system's performance under high thermal load. The longitude and latitude angels were designated to a location in Cairo, Egypt, at 15:00.

4.1 Baseline Simulation Setup

This initial test provides a benchmark for comparison between conventional HVAC operation and this new paradigm. In both the baseline and the EP model simulations, the Performance Index and heat map will be used to objectively compare the effectiveness of traditional versus dynamic control strategies.

The baseline simulation replicates a standard HVAC setup, where all terminals distribute cooled air uniformly across the office space. The amount of mass flow is controlled by a thermostat, that compares the average temperature with a temperature setpoint predetermined by typical building codes and comfort standards. In this scenario, the air distribution is even through all terminals.

Baseline Performance Index. Calculating the performance index of the conventional problem, temperatures from pixels of the surface were harvested. As per Eq. (1), the standard deviation of the temperatures from the comfort determined temperature 24 °C is 4.721 °C. With trial and error of mass-flow-inlet values, this was the least PI output. This indicates that temperature of most spots in plan, on average, deviate from 24 °C with a positive or a negative value of 4.721 °C, cooler or hotter.

Analysis of Baseline Heat Map. Initial study of the heat map from the baseline simulation shows temperature variances across the layout. Clear and significant temperature non-uniformity appears in plan, with different color zones illustrating different temperature zones. It is due to this non-uniformity that conventional HVAC controls fail to adjust to the dynamic thermal needs of diverse office areas Fig. 4. Due to direct exposure to sunlight, hot spots can be observed on the heat map, especially in proximity with the peripheral where large windows and glazing. Conversely, on the interior zones, cool spots appear at the absence of direct sunlight.

4.2 Implications of Findings

High PI, with high deviations from comfort temperature, and such variations in heat map analysis indicate that a uniform air distribution technique cannot appropriately capture the minor thermal intricacies within the plan. This discrepancy indicates inadequate thermal management and the emphasizes the challenges of sustaining uniform temperature across office layout.

5 Optimization

This section will demonstrate the effect of applying the proposed adaptive system to the same simulation case.

5.1 Implementation of the Novel Practice

Our approach begins with the integration of a Python script that seamlessly merges genetic algorithms with the 'Pyfluent API'. This integration enables dynamic simulations, promoting the systematic evaluation of different solutions iteratively.

The process starts with generating a population of solutions with a set of randomly generated mass-flow-inlet values at each VAV terminal. The script feeds the Ansys Fluent solver with each solution's values and runs the solver. At each Run, the Solver returns the temperature values of the predetermined surface as a 'csv file' to the script. Using Eq. (1) to calculate the PI, the fitness function of each solution is calculated, and the population of solutions are sorted accordingly, to keep them ranked based on their PI (those with the lowest index ranked highest). A selection of parents based in "Rank

Selection" method elects most fit solutions to transfer their genes-in our case the mass flow inlet values-to the next generation. This process is crucial to determine the optimum solution throughout the generations.

Crossover Process. Parents then undergo a crossover process that GA uses to combine the attributes of two selected parents. This combination of two parents produces two new children. Iteratively, this offsprings a new generation of ten solutions, each inheriting features from the parents, thereby electing the advantageous traits

Elitism and Mutation. To ensure diversity within the population and prevent the algorithm from becoming trapped in the local minima phenomena, a GA operator called 'Mutation' is introduced. This operator randomly, at a controlled rate, alters the genetic attributes of the new offsprings, allowing for more exploration of the solution space.

Concurrently, to prevent the high performing solutions from getting lost during exploration of search space, another operator called 'Elitism' is used. This operator carries the top solution over to the next generation in case no superior child emerged, safeguarding the most fit solution explored by the algorithm.

6 Results

This chapter presents the findings of the simulations conducted using the new paradigm. Followed by a comprehensive discussion of these results.

6.1 Analysis of Optimization of Performance Index

A graph plot of the optimization curve showcases the evolutive behavior of the genetic algorithm over successive generations. The PI of each generation is depicted as the least standard deviation of the 10 solutions in each generation (as shown in Fig. 3).

Initially, a steep descent is observed, which suggests a rapid change in the distribution of thermal comfort throughout the layout. This implies that the algorithm has quickly explored a wide range of possible solutions and promptly discovered a configuration that is close to the desired. As generations surpass the 20th generations, the descent starts to level off, implying that the algorithm is stuck in a local minima. However, fluctuation in the curve is observed, indicating that mutation is taking place and trying to escape the local minima. One of which has succeeded, a successful mutation occurred, altering one solution to be more fit, causing another descent in the values.

Reaching the 60th generation, the rate of improvement starts to dissipate, indicating that the algorithm is variating around the near-optimal values. Upon that state, the convergence pattern is almost complete, stabilizing the performance with only marginal improvements refining the solution. This pattern implies that further generations will have a low profit to computational consumption ratio. Which triggers the algorithm to stop iterating for an accepted near-optimal solution: a PI of 1.895 °C, 60% improvement from conventional PI.

6.2 Analysis of Optimized Heatmap

In view, comparing the two visual heat maps together, it is noticed that—according to the color map-a large portion of the plan is over cooled, orange reddish, in the conventional

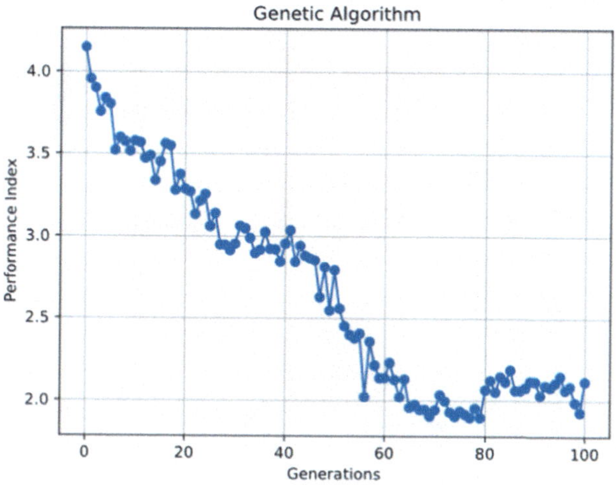

Fig. 3: Optimization Curve of Genetic Algorithm

method. Accordingly, the algorithm tries to balance the pixels to be approximately the same temperature. Therefore, the heat map of the optimized solution, in Fig. 5, shows a seamless color gradient with a surplus of the green-yellowish color, indicating the target temperature 24 °C, unlike the conventional solution heat map in Fig. 4.

However, although not all the spots in the optimized solution abide by the desired temperature, it is computationally exhausting to reach such optimum solution using current computation resources. This result, nevertheless, shows a promising potential for the effectiveness of such hypothesis.

7 Discussion

Results from performance index show that the new framework alleviates the temperature fluctuations, where most pixels stray away from the target comfort temperature with a 60% reduced value than conventional configuration. Vital to human comfort and the built environment, this improvement seems promising in optimizing the cooling of open space layouts, omitting over-cooling and over-heating. This framework has considered a human-centric approach in the cooling strategy. Based on the two heat maps, while the conventional configuration and optimized framework consume a total of 0.48 and 0.461 kg/s mass flow rate, respectively, the optimized framework shows optimal utilization of the consumed energy in space cooling.

This framework can be installed on BMS with an addition of dampers that control mass flow rates at each diffuser, to control the amount of cooled air exiting the diffuser. With such simple hardware modifications on building HVAC system and software update to BMS, it is possible to optimize the temperature of spaces dynamically to cope with the variating thermal loads. Moreover, this can enable dynamic partitioning of spaces, where layouts can vary in use based on function, and the system will adapt to these changes keeping the space in proximity to comfort levels.

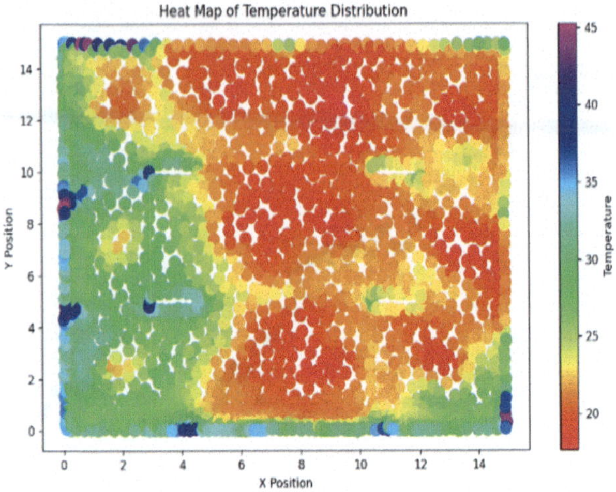

Fig. 4 Heat map of baseline solution.

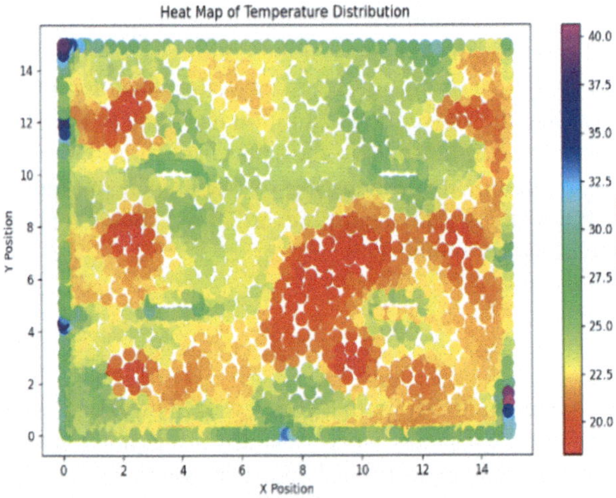

Fig. 5 Heat map of optimized solution.

8 Conclusion

Dealing with the HVAC system must be in regard with the architectural space, respecting the context locally. Due to ever changing nature of architectural spaces, thermal loads imposed from direct sunlight, occupancy, equipment, etc., the conventional approach of uniform cooling using thermal setpoints controlled be thermostats does not capture the thermal nuances in the plan, leaving some spots over-heated and other over-cooled. Hence the rising need to an adaptive approach, that treats each spot individually, but not independently, in regard with the overall thermal stability of the system.

This paper proposed an approach, referred to as the floating equilibrium point. This approach has been implemented and validated in this study quantitatively, using simulation and optimization. With results showing the effectiveness of utilizing VAV technology in a multi-agent system, that elects and adapts to the minor thermal fluctuations that occur inevitably. Furthermore, this work lines up with several Sustainable Development Goals (SDGs), specifically the ones focused on energy efficiency, sustainable cities and communities, and climate action. The proposed method contributes to reducing environmental impact, promoting sustainability, and ensuring human-centric building environments by enhancing the energy efficiency of HVAC systems.

References

1. Fikry, M.A.: Noise: simple attenuation tools. Architect. Plann. **18** (2007)
2. IEA, B.: Energy Use in Buildings. Paris (2022)
3. Vakiloroaya, V., Samali, B., Fakhar, A., Pishghadam, K.: A review of different strategies for HVAC energy saving. Energy Convers. Manag. **77**, 738–754 (2014). https://doi.org/10.1016/j.enconman.2013.10.023
4. Wu, D.C., Momeni, M., Razban, A., Chen, J.: Optimizing demand-controlled ventilation with thermal comfort and CO_2 concentrations using long short-term memory and genetic algorithm. Build Environ. **243** (2023). https://doi.org/10.1016/j.buildenv.2023.110676
5. Shi, H., Chen, Q.: Building energy management decision-making in the real world: A comparative study of HVAC cooling strategies. J. Build. Eng. **33** (2021). https://doi.org/10.1016/j.jobe.2020.101869
6. Shin, M., Haberl, J.S.: Thermal zoning for building HVAC design and energy simulation: a literature review. Energy Build. **203**. Elsevier Ltd (2019). https://doi.org/10.1016/j.enbuild.2019.109429
7. Bonilla, D., Samaniego, M.G., Ramos, R., Campbell, H.: Practical and low-cost monitoring tool for building energy management systems using virtual instrumentation. Sustain. Cities Soc. **39**, 155–162 (2018). https://doi.org/10.1016/j.scs.2018.02.009
8. Chou, J.S., Bui, D.K.: Modeling heating and cooling loads by artificial intelligence for energy-efficient building design. Energy Build **82**, 437–446 (2014). https://doi.org/10.1016/j.enbuild.2014.07.036
9. Seeam, A., Laurenson, D., Usmani, A.: Evaluating the potential of simulation assisted energy management systems: a case for electrical heating optimisation. Energy Build **174**, 579–586 (2018). https://doi.org/10.1016/j.enbuild.2018.06.063
10. Jiang, L., Yao, R., Liu, K., McCrindle, R.: An Epistemic-Deontic-Axiologic (EDA) agent-based energy management system in office buildings. Appl. Energy **205**, 440–452 (2017). https://doi.org/10.1016/j.apenergy.2017.07.081
11. Mariano-Hernández, D., Hernández-Callejo, L., Zorita-Lamadrid, A., Duque-Pérez, O., Santos García, F.: A review of strategies for building energy management system: Model predictive control, demand side management, optimization, and fault detect and diagnosis. J. Build. Eng. **33**. Elsevier Ltd (2021). https://doi.org/10.1016/j.jobe.2020.101692
12. Zhang, Y., Zeng, P., Zang, C.: Multi-objective optimal control algorithm for HVAC based on particle swarm optimization. In: Fifth International Conference on Intelligent Control and Information Processing, pp. 417–423 (2014). https://doi.org/10.1109/ICICIP.2014.7010290
13. Wang, F., Yoshida, H., Ono, E.: Methodology for optimizing the operation of heating/cooling plants with multi-heat-source equipments. Energy Build **41**(4), 416–425 (2009). https://doi.org/10.1016/j.enbuild.2008.11.012

14. Shaikh, P.H., Nor, N.B.M., Nallagownden, P., Elamvazuthi, I., Ibrahim, T.: A review on optimized control systems for building energy and comfort management of smart sustainable buildings. Renew. Sustain. Energy Rev. **34**, 409–429. Elsevier Ltd (2014). https://doi.org/10.1016/j.rser.2014.03.027
15. Garces-Jimenez, A., Gomez-Pulido, J.M., Gallego-Salvador, N., Garcia-Tejedor, A.J.: Genetic and swarm algorithms for optimizing the control of building hvac systems using real data: a comparative study. Mathematics **9**(18) (2021). https://doi.org/10.3390/math9182181
16. Aste, N., Manfren, M., Marenzi, G.: Building automation and control systems and performance optimization: a framework for analysis. Renew. Sustain. Energy Rev. 75, 313–330. Elsevier Ltd. (2017). https://doi.org/10.1016/j.rser.2016.10.072
17. Chi, F., Xu, Y.: Building performance optimization for university dormitory through integration of digital gene map into multi-objective genetic algorithm. Appl. Energy **307** (2022). https://doi.org/10.1016/j.apenergy.2021.118211
18. Yu, W., Li, B., Jia, H., Zhang, M., Wang, D.: Application of multi-objective genetic algorithm to optimize energy efficiency and thermal comfort in building design. Energy Build **88**, 135–143 (2015). https://doi.org/10.1016/j.enbuild.2014.11.063
19. Hosamo, H.H., Tingstveit, M.S., Nielsen, H.K., Svennevig, P.R., Svidt, K.: Multiobjective optimization of building energy consumption and thermal comfort based on integrated BIM framework with machine learning-NSGA II. Energy Build **277** (2022). https://doi.org/10.1016/j.enbuild.2022.112479
20. American Society of Heating, R., Engineers, A.-C.: ASHRAE Standard 90.1–2016: Energy Standard for Buildings Except Low-Rise Residential Buildings. ASHRAE, Atlanta, GA (2016)
21. McDowall, R.: Fundamentals of HVAC Systems. Elsevier, Oxford, UK (2007)
22. Wang, H., Chen, X., Vital, N., Duffy, E., Razi, A.: Energy optimization for HVAC systems in multi-VAV open offices: A deep reinforcement learning approach. Appl. Energy **356** (2024). https://doi.org/10.1016/j.apenergy.2023.122354
23. Nash, J.: Non-cooperative games. Ann. Math. **54**, 286–295 (1951)
24. Nasruddin, S., Satrio, P., Mahlia, T.M., Giannetti, N., Saito, K.: Optimization of HVAC system energy consumption in a building using artificial neural network and multi-objective genetic algorithm. Sustain. Energy Technol. Assess. **35**, 48–57 (2019). https://doi.org/10.1016/j.seta.2019.06.002
25. Singh Dhakar, P., Patel, A., Dhakar, P.S., Madhya Pradesh, G., Professor, A.: NITM Gwalior Madhya Pradesh, "Issue 11 www.jetir.org (ISSN-2349–5162) JETIR1811664 J. Emerg. Technol. Innov. Res. (JETIR) www.jetir," **5**, 436 (2018). https://doi.org/10.13140/RG.2.2.13462.50249
26. Afram, A., Janabi-Sharifi, F.: Review of modeling methods for HVAC systems. Appl. Thermal Eng. **67**(1–20), 507–519. Elsevier Ltd. (2014). https://doi.org/10.1016/j.applthermaleng.2014.03.055

Green Technology and Sustainable Development

Utilization of the Future Chart 2050 to Realize Sustainable Cities in Japan

Dipendra Dhakal(✉) and Hidefumi Kurasaka

Graduate School of Humanities and Social Sciences, Chiba University, Chiba, Japan
dipendha@gmail.com

Abstract. To realize a sustainable decarbonized society by 2050 amidst the adverse consequences of a declining and aging population, visions and policies must be designed to consider the perspectives of future generations. This study presents a framework that envisions a sustainable decarbonized society in 2050 by introducing a back casting approach to policy formulation. This study analyzes the "Future Chart 2050," a framework that simulates and numerically visualizes the future status of society across various sectors, including economic, social, environmental, and physical infrastructure. In this framework, policy changes in the current trend can change the future status indicated in the future chart. Based on the "Future Chart 2050," future workshops involving junior and senior high school students and young local government officials were conducted in various cities. Using the back casting policy approach, participants were given questionnaires to measure the effectiveness of their perspectives. This study reports on the policy measures proposed by the participants in future workshops. The findings revealed that the participants showed a strong interest in local issues, expressing a desire to contribute to the community, and feeling that they could collaborate with various individuals. They also showed enthusiasm for assuming future leadership roles and implementing activities aimed at improving local governance and addressing the problems of the host city. The recommendations realized from the workshop can be used for policies that anticipate the future of local governments, considering factors such as population, aging, industry, medical care, nursing care, and childcare, and can therefore support local governments in acting to achieve a sustainable decarbonized society by 2050.

Keywords: aging population · back casting · declining population · future chart · future workshop · sustainability

1 Introduction

Carbon neutrality has emerged as a prominent global trend, with over 120 countries and regions expressing their commitment to achieve it by 2050. Its goal represents an integrated and highly complex challenge intricately linked to climate, ecology, energy, the environment, and society [1]. The Japanese government pledged to attain carbon neutrality and achieve the societal goal of becoming a sustainable decarbonized society by 2050 by designing visions and policies that consider the perspectives of future generations [2].

G. Huang et al. (Eds.): ICESD 2024, ESE, pp. 125–143, 2025.
https://doi.org/10.1007/978-3-031-88683-6_10

However, realizing a sustainable decarbonized society has become a critical challenge both nationally and internationally, owing to the rising threat of climate change. To mitigate the impacts of climate change, the trend of global temperature rise, which began during the Industrial Revolution, must be limited to 1.5 °C. Achieving this target requires the worldwide attainment of net zero greenhouse gas (GHG) emissions by 2050 [3]. Consequently, countries leading decarbonization efforts must achieve carbon neutrality, and many countries have recently declared a goal to realize a decarbonized society by approximately 2050 [4]. In societies where the population is declining and aging over the long term, enduring challenges that surpass the temporal perspective of conventional administrative planning arise. This necessitates a departure from societies that rely on fossil fuels as an energy source to achieve carbon neutrality by 2050 [5].

This study systematically discusses the essential long-term responses aimed at achieving a sustainable decarbonized society. These encompass challenges extending beyond the temporal scope of conventional administrative planning, such as long-term population decline and aging, and the goal of carbon neutrality by 2050. Social investments must be allocated strategically to realize an ideal society. This involves addressing shortages in nursing care and medical services in urban areas, coping with the decline in the working-age population in rural areas, promoting compact living arrangements, formulating selective settlement strategies, transitioning from centralized to distributed energy supply systems, implementing zero-energy building (ZEB) and zero-energy house (ZEH) concepts in construction, and sustainably investing in renewable energy sources.

Japan's population peaked in 2008 and has since entered a period of overall decline that is expected to continue until the year 2050, with the elderly population reaching its maximum by 2040. As illustrated in Fig. 1, Japan's total population is expected to reach 106.42 million by 2045, representing a decrease of more than 20 million compared with that in 2015. This decline is equivalent to the loss of the entire population of the Tokyo and Chiba prefectures. In Japan, the aging population and depopulation in cities have resulted in significant challenges, including a decline in community vitality in most secondary cities [6]. In a society with a declining and aging population, the shortage of human resources in each sector supporting sustainability is expected to become more serious [7].

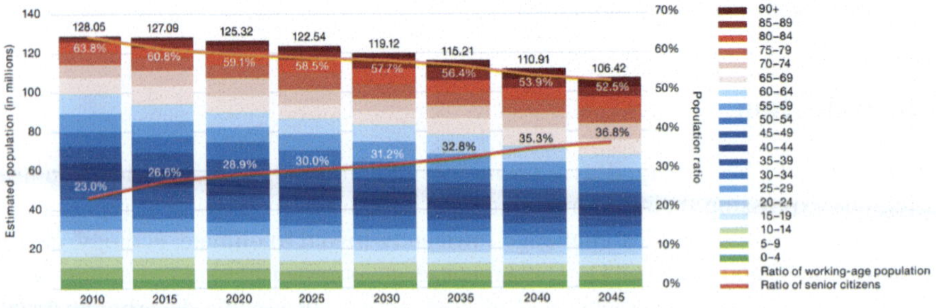

Fig. 1. Demographic changes in Japan. *Source* IPSS, 2018

As the central hubs of economic growth, cities are the world's major consumers of energy resources, consuming 67% of the global energy demand, and are significant contributors to GHG emissions, generating up to 70% of harmful GHG emissions [8]. Sustainable cities have faced criticism for encountering challenges in translating sustainability into the built environment and evaluating the extent to which various sustainable urban forms contribute to the goals of sustainable development [9–13]. Social investments must be made systematically to realize an ideal society, such as responding to the shortage of nursing care and medical services in urban areas, the decline in the working-age population in rural areas, compact ways of living, selective settlement, a shift from centralized energy supply to distributed energy supply, ZEB/ZEH, and sustainable investment in renewable energy.

Dreborg [14] mentioned that forecasting involves predicting the future by monitoring and analyzing current trends and making projections based on explicit or implicit assumptions derived from present observations. Essentially, it initiates the planning process based on the current situation and generalizes today's trends and practical solutions for the future. Therefore, all stakeholders should collectively predict the kind of society they desire to create from the temporal perspectives of 2040 and 2050 in order to consider the ideal future. This requires the cooperation of the members of society to create an ideal image.

Future generations should be important stakeholders in any discussion on planning for society until 2050. However, in reality, they are unable to participate in current debates and decision-making. Given the current social system, decision making that explicitly considers the interests of future generations is difficult. For this reason, back casting, as a new approach, is well suited to multifaceted planning processes [15] to describe the images of society in 2050 and formulate the measures needed to realize it in a manner that allows the consideration of future generations' perspectives. However, only a few studies have investigated the realization of a future society using future charts. To address this gap, this study aimed to create an image of society using the Future Chart, which predicts the population and age structure of each city in 2050 based on the population projections (median estimates) of the National Institute of Population and Social Security Research. Consequently, stakeholders must reach a consensus on how to address specific technical, social, and economic challenges unique to each region to cover decarbonization and sustainability [16, 17]. Therefore, regions and local governments must establish their own goals and implement measures to address sustainability and decarbonization goals.

In 2021, Japan revised its Global Warming Promotion Act to focus on "the realization of a decarbonized society by 2050 in Japan." Furthermore, Japan's Nationally Determined Contribution submitted to the Paris Agreement Secretariat in October 2021 set the carbon neutrality target for 2050 and specifies that "we aim to reduce greenhouse gas emissions by 46% in FY2030 compared to FY2013 and continue to challenge ourselves to achieve a 50% increase." According to the Japanese government, Ministry of the Environment, by the end of January 2022, 534 municipalities (including 40 prefectures, 319 cities, 15 special districts, 134 towns, and 26 villages) had declared their commitment to achieve decarbonization by 2050. Additionally, they plan to establish at least 100 leading decarbonization regions by 2030. To simultaneously address the realization of a decarbonized society and resolve regional issues, achieving a compact city based on the future population size and establishing a regional circulating and ecological sphere that utilizes the potential of local resources is necessary.

This study pioneers in conducting participatory workshops with junior high school and high school students using a future chart. Future Charts predict the future images of cities based on current statistics and policies if no change occurs in the current situation. This prediction helps in understanding the desired future while highlighting the gap between the current and future situations (Fig. 2). However, considering policy (back casting) for solutions can change the predicted image. Back casting involves starting with a desirable future vision and then working backwards to the present to identify the most strategic steps or actions necessary to achieve a projected future state for each city by 2050. This approach considers the impact of a declining population and an aging society in various municipal sectors.

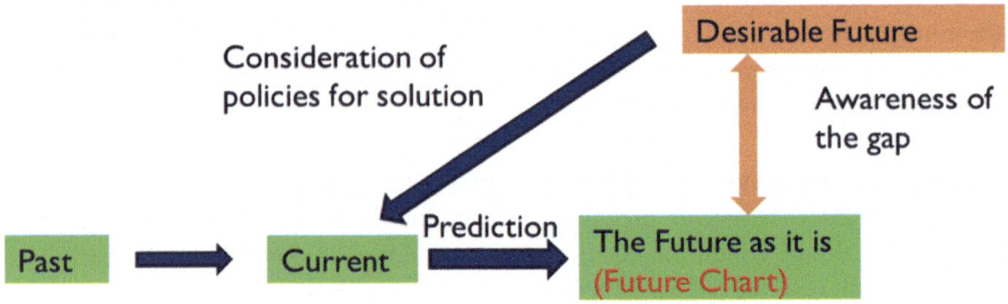

Fig. 2. Use of a Future Chart. *Source* Authors

Thus, this study represents a groundbreaking example of policy design by integrating a new mechanism that combines future charts of municipalities with systems thinking. This integration supports the objectives outlined in the Paris Agreement to mitigate global warming, which is necessary to realize a sustainably decarbonized society by the year 2050.

2 Methods

2.1 Content and Use of Future Chart

The future chart visualizes the future status if the current situation continues in various areas, including population, industrial structure, childcare and education, medical and nursing care, public facilities, roads, housing, farmland and agriculture, fisheries, forests, forestry, renewable energy, and public finances. By simply entering the municipality code, the future chart records of 1,741 municipalities nationwide were issued.

In Fig. 3, we have presented only the flow of the content of the future chart which is described how to create a future chart in detail in the footnote below the figure.

Figure 3(1): For the age groups in 5-year intervals for individuals aged 25 and above, the method involves calculating the ratio of the number of employees in a certain industry

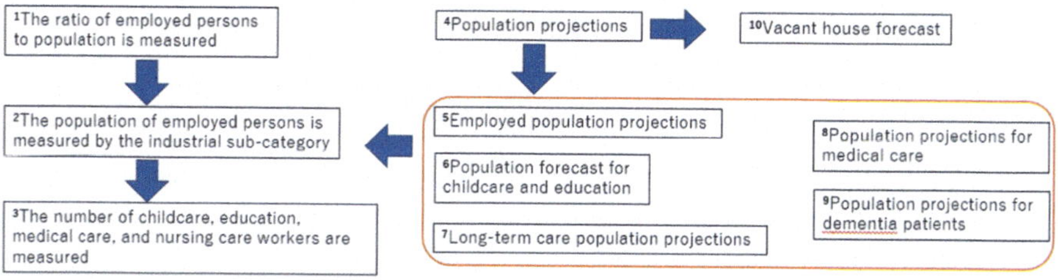

Fig. 3. Content of a Future Chart. *Source* Authors

(broad classification) for each census year (2000, 2005, 2010, 2015, 2020) divided by the number of employees in the same industry for the previous census year. This ratio is calculated for four pairs of consecutive census years (2000/2005, 2005/2010, 2010/2015, 2015/2020), and their average is calculated. This average ratio is then multiplied by the number of employees in the specified industry for the (α) age group in the 2020 census. This multiplication yields an estimate of the number of employees in the $\alpha + 5$ age group for the specified industry in the year 2025. This process is repeated until the year 2050. For estimating the employment-population for the 15–19 and 20–24 age groups beyond 2025, the method involves multiplying the number of employees in these age groups in 2020 by the ratio of the number of employees for the corresponding age groups in 2020 to the total employment population for the 15–19 and 20–24 age groups in 2020 across all industries. This methodology is applied to estimate the employment population for different industries categorized by broad classifications for each municipality in the years 2025, 2030, 2035, 2040, 2045, and 2050.

Figure 3(2): By multiplying the projected employment population by the projected ratio of the employed population by major industry category, the employed population by major industry category was predicted.

Figure 3(3): The 2020 employment population (by the municipality) of childcare workers, teachers, doctors, and nursing care workers in the corresponding major industry category was fixed by dividing the number of people by each municipality in 2020 and estimated by multiplying it by 2025 and beyond.

Figure 3(4): For population estimates by municipality up to 2045, data regarding gender, age (5 years), and class were utilized from the "Future Population Projections by Region in Japan (2018 estimated)" provided by the National Institute of Population and Social Security Research. In the case of Fukushima Prefecture, population projections by the municipality were extended up to 2035 using forecasts for the entire prefecture, and subsequently extrapolated until 2045. Population estimates for 2050 were derived by extending and consolidating the population trends projected for each municipality, disaggregated by 5-year age groups for both men and women, up to the year 2045.

Figure 3(5): The employed population for the years 2025, 2030, 2035, 2040, 2045, and 2050 was forecasted by multiplying the population projections for men and women in the 5-year age group category by the number of employed individuals within the same age group and gender in the respective municipality in 2020. This multiplication was carried out alongside the population figures for men and women in the 5-year age group category obtained from the 2020 census.

Figure 3(6): Using the population projections for males and females in the 5 year age group, the population aged 0–4 years old was combined with the population aged 5–9 years old by adding 2/5 of the former to calculate the population eligible for childcare. The elementary school population was determined by adding 3/5 of the population aged 5–9 years old and 3/5 of the population aged 10–14 years old. The junior high school population was calculated by adding 2/5 of the population aged 10–14 years old and 1/5 of the population aged 15–19 years old.
Figure 3(7): In the "Long-term Care Insurance Business Status Report" (by insurer, as of the end of February 3, Reiwa 3), the ratio of support required and long-term care required for each municipality by gender in the 65-to-5 age category was calculated (municipalities that implement long-term care insurance in some administrative associations apply the ratio of some administrative associations to which they belong), and this ratio was applied to the population forecast of each municipality to predict the population eligible for long-term care after 2025.
Figure 3(8): Utilizing the "Overview of Patient Surveys" since 1999 and "Statistical Table of Medical Treatment Rates (per 100,000 Population) Hospitalizations by Outpatient, Yearly, Sex, and Age Groups," we forecasted future hospitalization and outpatient rates. This was achieved by logarithmically extrapolating the hospitalization and outpatient rates for each 5 year age group among males and females in the prefecture to which each municipality belongs. Subsequently, these rates were applied to the population projections of each municipality for individuals aged 65-to-5-years-old, categorized by gender, beyond the year 2025.
Figure 3(9): Supplementary Appendix which outlines the prevalence of dementia by sex and age group (%), was derived from a mathematical model. This model was built on the assumption of a 20% increase in the frequency of diabetes from 2012 to 2060. The data originates from the "Comprehensive Research Report on Future Estimation of the Elderly Population with Dementia in Japan" (2015), led by Principal Investigator Toshiharu Ninomiya, as part of the Grant-in-Aid for Health, Labour and Welfare Science (MHLW Science Special Research Project). The proportion of dementia patients in the 65-to-5-year-old category, categorized by gender, was estimated by applying these findings to the population forecast of the 65-to-5-year-old category, also categorized by gender, in each municipality after 2025.
Figure 3(10): The logarithmic return results of the number of individuals per household for each prefecture are applied to the population forecast of each municipality to predict the future number of households in that area. Drawing from the 2018 Land and Housing Statistics, the floor area of houses per household is calculated and applied to the forecasted number of households to estimate the required floor area in the future. Conversely, for existing houses, those constructed between 1971 and 1980 will experience a depreciation rate of 0.6, while those built before 1970 will have a depreciation rate of 0.8. Furthermore, utilizing statistics on new construction starts, the floor area of newly constructed houses is estimated. This approach helps in estimating the residential floor area that will be present in the future. By subtracting the required floor area from the existing floor area, an estimation of the floor area that will be vacant is obtained.

2.2 Framework and Setting of Future Workshops

This study was conducted through a participatory workshop held in collaboration with the host cities. The future workshop employed future charts to involve junior and senior high school students, as well as young professionals, who will shape the future of the region and potential scenarios for 2050 if the current trajectory persists. The participants were encouraged to anticipate policy proposals from respective city officials. For each municipality, various graphs were used to depict potential societal outcomes that should continue until 2050, thereby enhancing awareness through predictive perceptions.

The workshop sessions were held in schools or city hall buildings, with the duration varying depending on the participants and content, including half-day, full-day, and two-day courses. All the sessions were recorded for subsequent analysis. The participants were divided into groups based on their numbers, and the members of each group remained the same throughout the workshop sessions. Throughout the sessions, participants examined policies and measures aimed at shaping a sustainable decarbonized society from the perspective of future generations. Figure 4 shows an example of a workshop session.

Fig. 4. Snapshot of a Future Workshop. *Photo* Authors

To facilitate the sessions, each group appointed one member as a facilitator, whose role was to encourage participants to explore ideas related to the areas covered in future charts. Subsequently, the facilitator summarizes and visualizes the opinions expressed by the group members. Figure 5 illustrates the flow and framework of the workshop, which was divided into five sessions. Details of the discussion content and design are provided below along with an explanation of each session.

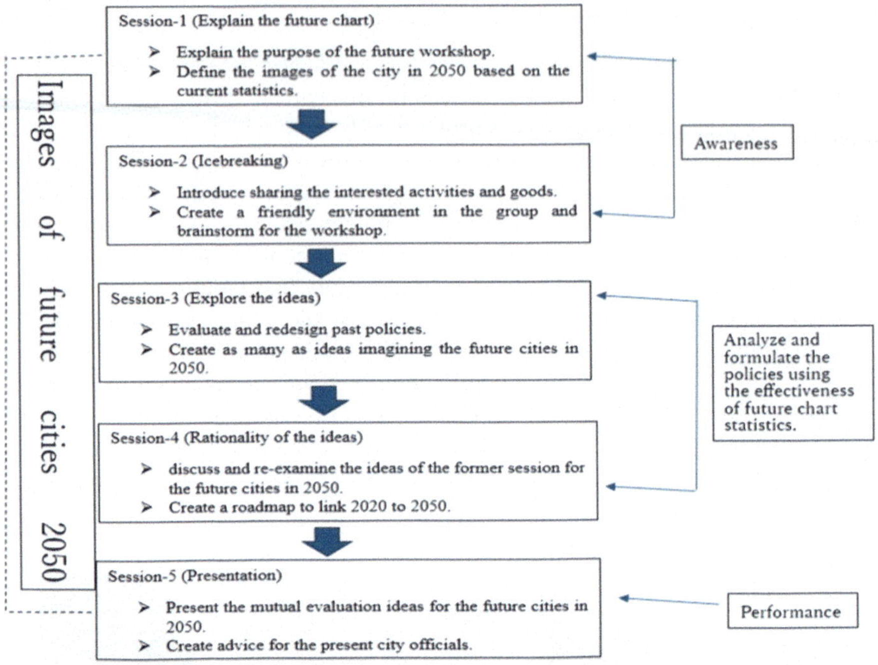

Fig. 5. Flow and Framework of the Workshop. *Source* Authors

Session 1: During the initial session, after the assembly of participants, the general facilitator explained the workshop's objectives and provided an overview of the future chart records, sharing information on the status of the population, industrial structure, childcare care and education, medical and nursing care, public facilities, roads, housing, farmland and agriculture and fisheries, forests and forestry, renewable energy, global warming, disasters, technological progress, and public finances. The participants listened to the lecture while referring to their workbooks.

Session 2: Participants engaged in an icebreaking activity, in which they wrote their names in the center of a paper and surrounded it with words related to their interests, activities, foods, goods, and so on. Following this, participants rotated the naming paper clockwise and asked neighboring individuals questions based on the words written in their respective papers, facilitating mutual understanding and interaction among participants.

Session 3: Participants were tasked with generating ideas individually across various fields, envisioning a future city in 2050 based on statistics from the future chart. Each participant was instructed to write one idea per piece of sticky paper. Upon completing their ideas, the participants engaged in group discussions to share and categorize their ideas based on their respective fields. They then attached sticky notes to half of a cardboard paper, grouped ideas from the same field, and labelled each section accordingly.

Session 4: Participants were prompted to critically review and re-evaluate their collective societal vision for 2050. They engaged in group discussions to justify their ideas, drawing on insights from the previous session (Session 3) regarding the projected societal conditions in 2050. Subsequently, the participants attached their rationalized ideas to the other side of the cardboard, grouped them, and reiterated the names of the fields. Following this, all participants, facilitators, and guests (optional) were provided with two types of small stickers, called "*iine shiru*" (approval stickers), with one set in red for participants and another in blue for facilitators and guests. Anyone among all groups was encouraged to freely issue stickers, indicating the approval of ideas from their respective perspectives.

Session 5: The participants were tasked with presenting their best ideas collectively within their respective groups. The ideas that received the most stickers were considered the best and most significant, although all ideas were considered valuable for formulating future design policies (futurability) from the perspective of individuals living in 2050. Following the participants' presentations, the guest (mayor or their representative) was invited to conclude the presentation based on the ideas presented by the participants. Support ideas for the current mayor to formulate policies in the present context were also welcomed. Notably, throughout Sessions 3–5, the participants were encouraged to engage in discussions and explore ideas from the perspective of individuals living in 2050.

2.3 Questionnaire Survey Analysis for Effectiveness of Participant Questionnaires

After each workshop, a questionnaire survey was administered to all participants to evaluate how their perspectives and attitudes evolved as a result of the adopted framework and treatments. Participants were presented with multiple items and asked to indicate their level of agreement with statements such as, "As a result of participating in the workshop, I feel this way more than before." The questionnaire covered the following areas: (1) changes in their perception of the relationship between current and future generations, (2) shifts in their evaluation of the host city as a town, and (3) changes in their perception of important aspects of policymaking.

3 Results and Discussion

3.1 Measuring the Effectiveness of Participant Questionnaires

This section examines the effectiveness of the questionnaire survey following a future workshop. As outlined in Fig. 5, the workshop involved awareness of the current and future statistics of the host city if it remained on the same trajectory (Sessions 1 and 2), analyzing and formulating policies using the effectiveness of future chart statistics (Sessions 3 and 4), and mutual performance by participants envisioning future cities by 2050 (Session 5). Building upon the framework of future workshop policies, this study particularly focuses on the insights garnered from Sessions 3 and 4, as these are expected to have a direct impact on the activation of futurability.

The cities where future workshops were conducted are indicated on Japan's political map (Fig. 6). The red points denote the cities and buildings where future workshops take place, along with the corresponding dates.

After the workshop, participants were given the options outlined in Table 1 as a questionnaire survey and were asked to respond to multiple items, such as "As a result of participating in the workshop, I feel this way more than before."

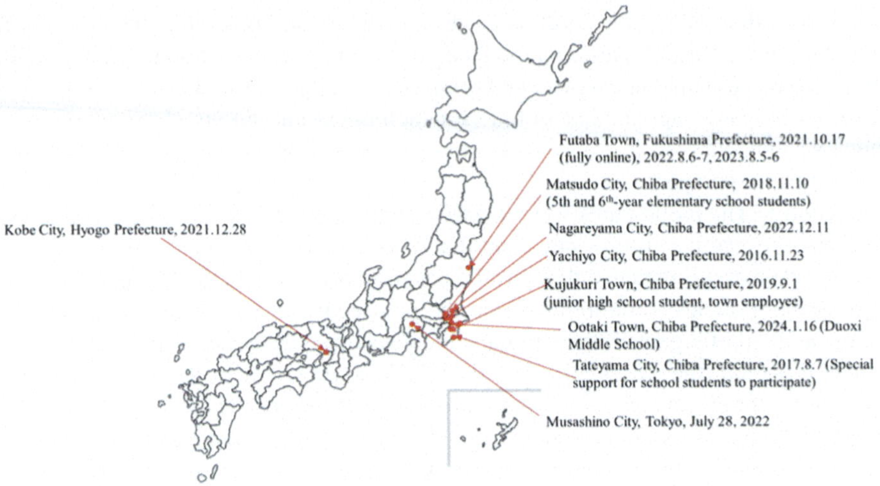

Fig. 6. Future Workshop Locations until January 2024. *Source* Authors

Table 1. Percentage of participants who selected each item

Options given to the participants	Percentage (%) of Participants' Responses by Cities								
	Yachiyo City	Tateyama City	Matsudo City	Kujukuri Town	Ootaki Town	Kobe City	Musashino City	Futaba Town	Nagareyama City
I like the host municipality	45%	82%	82%	33%	53%	79%	61%	57%	55%
I want to contribute in any way I can to make the host municipality a place easy to live	75%	89%	86%	75%	58%	95%	78%	81%	80%
I would like to know more about the problems in the host municipality	70%	70%	77%	50%	53%	89%	89%	86%	60%
I want to talk more about the problems that are happening in society and the community with my family and friends	40%	57%	55%	25%	26%	63%	72%	57%	60%

(*continued*)

Table 1. *(continued)*

Options given to the participants	Percentage (%) of Participants' Responses by Cities								
	Yachiyo City	Tateyama City	Matsudo City	Kujukuri Town	Ootaki Town	Kobe City	Musashino City	Futaba Town	Nagareyama City
By getting involved, I may be able to change the host municipality in a desirable direction	60%	36%	41%	25%	26%	16%	67%	19%	35%
If I am actively involved, I can influence the decision of the host municipality	50%	39%	36%	33%	21%	5%	61%	19%	40%
I would like to participate in a place where we can discuss together to solve problems that arise in the local government	45%	61%	64%	42%	26%	84%	67%	62%	45%
I would like to create a place where stakeholders can discuss and cooperate on the issues of the host municipality	40%	50%	55%	42%	26%	47%	61%	57%	45%
We would like to participate in activities to convey our opinions to the relevant parties on the issues of the host municipality	40%	68%	45%	33%	32%	84%	44%	43%	40%
Now, or when I grow up, I want to be a leader in activities that improve the local government where the event is held	20%	25%	27%	17%	16%	26%	22%	14%	15%

(continued)

Table 1. (*continued*)

Options given to the participants	Percentage (%) of Participants' Responses by Cities								
	Yachiyo City	Tateyama City	Matsudo City	Kujukuri Town	Ootaki Town	Kobe City	Musashino City	Futaba Town	Nagareyama City
I am knowledgeable about the current and future issues of the host municipality	50%	39%	45%	33%	32%	26%	39%	57%	45%
You will be able to think about the problems that are happening now in the host municipality from a long-term perspective	50%	55%	64%	33%	26%	47%	56%	71%	35%
You can think about how to prevent problems that are likely to occur in the future	55%	52%	64%	42%	37%	58%	61%	52%	50%
Able to come up with effective strategies that are highly viable	50%	52%	45%	33%	37%	53%	33%	62%	45%
You can come up with strategies that involve a variety of people with different positions and opinions	25%	48%	45%	33%	26%	53%	39%	57%	40%
I can work well as a team by collaborating with others	50%	55%	73%	33%	37%	89%	44%	81%	45%
Even people with different positions and opinions can work together to solve problems	65%	70%	77%	42%	53%	84%	50%	90%	45%

(*continued*)

Table 1. (*continued*)

Options given to the participants	Percentage (%) of Participants' Responses by Cities								
	Yachiyo City	Tateyama City	Matsudo City	Kujukuri Town	Ootaki Town	Kobe City	Musashino City	Futaba Town	Nagareyama City
In order to solve the problems of the host municipality, it is more effective to cooperate with others than to act individually	–	82%	82%	58%	79%	100%	67%	100%	65%
If we don't work together with others, the host municipality will be in a critical situation	–	59%	68%	67%	68%	74%	50%	67%	70%
It's our responsibility to work with others to tackle local issues together	–	61%	73%	33%	53%	89%	61%	76%	70%
Answers	19	43	22	12	19	19	18	21	20

Source Future Workshop until January, 2024

Table 1 shows the percentage of participants who selected each item from the options provided. The results presented in Table 1 are interconnected with those in Table 2, where the percentages are ranked in descending order to facilitate readers' comprehension.

Table 2 presents various options for the host city to consider in becoming a sustainable city by 2050. The image of the host city in 2050, developed by the participants in each group acting as future officials of the city, shared many common features. Some of the top-ranking features are highlighted in the yellow-colored columns, such as "I want to contribute in any way I can to make (the host city) a place where it is easy to live," "I would like to know more about the problems that are occurring in the host city," and "In order to solve the problems of the host city, it is more effective to cooperate with others than to act individually." In contrast, the lowest-ranking feature is mentioned in the brown-colored column, which states "Now, or when I grow up, I want to be a leader in activities that improve (the local government where the event is held)."

Across all future workshops, a consistent observation was that participants tended to develop an interest in local issues, expressed a desire to contribute to the community, and were motivated to collaborate with various individuals. The future workshop could potentially serve as an effective tool for civic education, particularly with the lowering of the voting age, and as a component of comprehensive study time, which has gained importance with the revision of junior high and high school curricula. However, arousing the desire to become a leader in the region through this workshop alone is challenging.

Table 2. Ranking of the percentage of participants who selected each item (in descending order)

Options given to the participants	Ranking of Participants' Responses by Cities								
	Yachiyo City	Tateyama City	Matsudo City	Kujukuri Town	Ootaki Town	Kobe City	Musashino City	Futaba Town	Nagareyama City
I like the host municipality	11	2	2	9	4	9	7	11	7
I want to contribute in any way I can to make the host municipality a place easy to live	1	1	1	1	3	2	2	4	1
I would like to know more about the problems in the host municipality	2	4	4	4	4	3	1	3	5
I want to talk more about the problems that are happening in society and the community with my family and friends	13	10	12	18	13	11	3	11	5
By getting involved, I may be able to change the host municipality in a desirable direction	4	19	18	18	13	19	4	18	18
If I am actively involved, I can influence the decision of the host municipality	6	17	19	9	19	20	7	18	15
I would like to participate in a place where we can discuss together to solve problems that arise in the local government	11	7	9	5	13	6	4	9	9

(*continued*)

Table 2. (*continued*)

Options given to the participants	Ranking of Participants' Responses by Cities								
	Yachiyo City	Tateyama City	Matsudo City	Kujukuri Town	Ootaki Town	Kobe City	Musashino City	Futaba Town	Nagareyama City
I would like to create a place where stakeholders can discuss and cooperate on the issues of the host municipality	13	15	12	5	13	15	7	11	9
We would like to participate in activities to convey our opinions to the relevant parties on the issues of the host municipality	13	6	14	9	11	6	15	17	15
Now, or when I grow up, I want to be a leader in activities that improve the local government where the event is held	17	20	20	20	20	17	20	20	20
I am knowledgeable about the current and future issues of the host municipality	6	17	14	9	11	17	17	11	9
You will be able to think about the problems that are happening now in the host municipality from a long-term perspective	6	11	9	9	13	15	12	7	18
You can think about how to prevent problems that are likely to occur in the future	5	13	9	5	8	12	7	16	8

(*continued*)

Table 2. (*continued*)

Options given to the participants	Ranking of Participants' Responses by Cities								
	Yachiyo City	Tateyama City	Matsudo City	Kujukuri Town	Ootaki Town	Kobe City	Musashino City	Futaba Town	Nagareyama City
Able to come up with effective strategies that are highly viable	6	13	14	9	8	13	19	9	9
You can come up with strategies that involve a variety of people with different positions and opinions	16	16	14	9	13	13	17	11	15
I can work well as a team by collaborating with others	6	11	6	9	8	3	15	4	9
Even people with different positions and opinions can work together to solve problems	3	4	4	5	4	6	13	2	9
In order to solve the problems of the host municipality, it is more effective to cooperate with others than to act individually	–	2	2	3	1	1	4	1	4
If we don't work together with others, the host municipality will be in a critical situation	–	9	8	2	2	10	13	8	2
It's our responsibility to work with others to tackle local issues together	–	7	6	9	4	3	7	6	2

Source Future Workshop until January, 2024

Future workshop sessions proved to be highly engaging in terms of both quality and content, surpassing the forecasts provided by future charts formulated from the perspective of current statistics. Consequently, the workshop aimed to foster the adoption of a systemic approach by considering futurability and proposing policies for new mechanisms and systems that do not yet exist. The response options for the questionnaire survey were developed throughout the future workshop, considering the activities of younger generations for the future society in 2050.

Table 3. Effectiveness of participant questionnaires

I want to contribute in any way I can to make (the host municipality) a place where it is easy to live
I would like to know more about the problems that are occurring in the host municipality
In order to solve the problems of the host municipality, it is more effective to cooperate with others than to act individually
Now, or when I grow up, I want to be a leader in activities that improve (the local government where the event is held)
I am knowledgeable about the current and future issues of the host municipality

Source Authors

Table 3 summarizes the effectiveness of the participants' questionnaires completed after the workshops. Participants analyzed the impact and effectiveness of each item on discussion and decision making (responses received). Findings from the questionnaire survey demonstrated that discussions from the future workshops can support officials and residents of host cities in the development of a greater understanding of the goal of a sustainable and decarbonized society, increased awareness of future challenges, and a heightened sense of hope for the future city by 2050.

This study also examined the effectiveness of utilizing future charts for discussions and defined the degree of liveliness, divergence, and development of discussions as indicators of discussion activity. The results of this study indicate that projections of socioeconomic status are effective in workshop discussions for increasing liveliness, reducing divergence, and enhancing the development of important themes. These effects were confirmed using questionnaire surveys. Moreover, systematizing the workshop flow is essential for the framework to be readily applied to policy design at any level for the achievement of a sustainable and decarbonized society by 2050.

4 Conclusion

The realization of a sustainable decarbonized society by 2050 is a societal goal that cannot be achieved solely by extending current systems, statuses, and ways of thinking. To effectively achieve social transformation toward this goal, visions and policies that consider the perspectives of young generations, while also encouraging, must be designed. This study proposes an innovative framework for envisioning a sustainable decarbonized society by 2050 and designing policies by integrating back casting and forecasting systems thinking. Participants formulated future images of the host city in a workshop and verified the effectiveness of the questionnaires (Table 3), which could support city officials in fostering futurability. This study demonstrated the effectiveness of the framework in systematically identifying the images of society in 2050 and the policy measures required to activate futurability among young generations.

Overall, the workshop participants tended to develop an interest in local issues, contribute to the community, and feel capable of collaborating with various people. However, the challenge of arousing the desire to become a leader in the region remains. The future workshop may prove effective as a tool for civic education, especially given the lowering of the voting age, and as part of a comprehensive study time, which has gained significance with the revision of junior high and high school curricula. In the future, the authors aim to propose more effective methods for promoting a decarbonized society. Future studies should explore the possibility of generalizing the findings by conducting additional case studies on the effectiveness of chart workshops and other systems thinking tools.

Acknowledgements. This work was supported by the Research Center for 'Co-JUNKAN' platform towards Beyond "Zero-Carbon" (JPMJPF2003), which is established by Japan Science and Technology Agency (JST) under the program on open innovation platform for industry-academia cocreation (COI-NEXT).

References

1. Huang, M.T., Zhai, P.M.: Achieving the Paris Agreement temperature goals requires carbon neutrality by the middle century with far-reaching transitions in the whole society. Adv. Clim. Chang. Res. **12**(2), 281–286 (2021)
2. Nakai, T.: Japan's triple sustainability challenge. Asia Pac. Bus. Rev. 1–15 (2024). https://doi.org/10.1080/13602381.2024.2320541
3. IPCC.: Global warming of 1.5 °C. An IPCC Special Report on the impacts of global warming of 1.5 °C above pre-industrial levels and related global greenhouse gas emission pathways, in the context of strengthening the global response to the threat of climate change (2018)
4. Rockström, J., Gaffney, O., Rogelj, J., Meinshausen, M., Nakicenovic, N., Schellnhuber, H.J.: A roadmap for rapid decarbonization. Science **355**(6331), 1269–1271 (2017). https://doi.org/10.1126/science.aah3443
5. Kudo, S., Mutisya, E., Nagao, M.: Population aging: an emerging research agenda for sustainable development. Soc. Sci. **4**(4), 940–966 (2015). https://doi.org/10.3390/socsci4040940
6. Lee, J., Newman, G.: Forecasting urban vacancy dynamics in a shrinking city: a land transformation model. ISPRS Int. J. Geo Inf. **6**(4), 124 (2017). https://doi.org/10.3390/ijgi6040124
7. Wang, Y., Fukuda, H.: Sustainable urban regeneration for shrinking cities: a case from Japan. Sustainability **11**(5), 1505 (2019). https://doi.org/10.3390/su11051505
8. Creutzig, F., Baiocchi, G., Bierkandt, R., Pichler, P.P., Seto, K.C.: A global typology of urban energy use and potentials for an urbanization mitigation wedge. Proc. Natl. Acad. Sci. **112**, 6283–6288 (2015). https://doi.org/10.1073/pnas.1315545112

9. Bibri, S.E., Krogstie, J.: ICT of the new wave of computing for sustainable urban forms: their big data and context-aware augmented typologies and design concepts. Sustain. Cities Soc. **32**, 449–474 (2017). https://doi.org/10.1016/j.scs.2017.04.012
10. Hofstad, H.: Compact city development: high ideals and emerging practices. Eur. J. Spatial Dev. **10**(5), 1–23 (2012)
11. Jabareen, Y.R.: Sustainable urban forms: their typologies, models, and concepts. J. Plan. Educ. Res. **26**, 38–52 (2006). https://doi.org/10.1177/0739456X05285119
12. Kärrholm, M.: The scaling of sustainable urban form: some scale—related problems in the context of a Swedish urban landscape. Eur. Plan. Stud. **19**(1), 97–112 (2011)
13. Williams, K.: Sustainable cities: research and practice challenges. Int. J. Urban Sustain. Dev. **1**(1), 128–132 (2009). https://doi.org/10.1080/19463131003654863
14. Dreborg, K.H.: Essence of back casting. Futures **28**(9), 813–828 (1996). https://doi.org/10.1016/S0016-3287(96)00044-4
15. Phdungsilp, A.: Futures studies' back casting method used for strategic sustainable city planning. Futures **43**(7), 707–714 (2011). https://doi.org/10.1016/j.futures.2011.05.012
16. Geels, F.W., Sovacool, B.K., Schwanen, T., Sorrell, S.: Sociotechnical transitions for deep decarbonization. Science **357**(6357), 1242–1244 (2017). https://doi.org/10.1126/science.aao3760
17. Urrutia-Azcona, K., Tatar, M., Molina-Costa, P., Flores-Abascal, I.: Cities4ZERO: overcoming carbon lock-in in municipalities through smart urban transformation processes. Sustainability **12**(9), 3590 (2020). https://doi.org/10.3390/su12093590

Modeling of New Energy Vehicles' Impact on Urban Ecology Focusing on Behavior

Run-Xuan Tang(✉)

Beijing Jiaotong University, Beijing, China
22241101@bjtu.edu.cn

Abstract. The surging demand for new energy vehicles is driven by the imperative to conserve energy, reduce emissions, and enhance the ecological ambiance. By conducting behavioral analysis and mining usage patterns of new energy vehicles, particular patterns can be identified. For instance, overloading the battery, operating with low battery power, and driving at excessive speeds can all detrimentally affect the battery's performance. To assess the impact of such driving behavior on the urban ecology, an environmental computational modeling method has been proposed to simulate the interaction between new energy vehicles and the environment. To extend the time series data of the vehicle's entire life cycle and the ecological environment within the model sequence data, LSTM model with Bayesian optimizer is utilized for simulation. The analysis revealed the detrimental effects of poor driving behavior on the environment.

Keywords: New energy vehicles · Environmental computing modeling · Behavior analysis · Battery · Deep learning

1 Introduction

In promoting new energy vehicles and hybrids, while the 19% global market share in 2022 positively impacts environmental protection and diminishes fossil fuel consumption, concerns regarding the battery's environmental effects are increasingly prevalent and represent a significant environmental challenge [1] The technological restrictions generated by batteries encompass vital environmental issues such as the strain exerted on the grid by excessive energy demand [2] and battery material recycling [3].

Mousaei [4] argues that the management of battery temperature must not be overlooked when evaluating energy management for hybrid electric vehicles (HEVs), as it significantly affects battery longevity. The Sitapure and Kulkarni study [5] highlights the importance of predicting the state of the battery, including charge and temperature, in creating an advanced battery management system (BMS).

Additionally, literature indicates that data-driven time series modeling was developed to predict battery lifetime, showing the influence of environmental factors and integrated parameters on battery performance. Cui et al. [6] conducted a study comparing new methods for EV charging and battery energy conditioning. They showed that enhancing the power handling system of aging battery packs can significantly improve

G. Huang et al. (Eds.): ICESD 2024, ESE, pp. 144–155, 2025.
https://doi.org/10.1007/978-3-031-88683-6_11

the performance of a used battery storage system (2-BESS). Another study by Alexeenko and Charisopoulos [6] demonstrated that utilizing a shared battery resource can substantially reduce the deployed battery capacity, which has the potential to improve battery efficiency.

For the issue of driving behavior in new energy vehicles, Gebbran et al. [7] highlighted the significance of minimizing losses and optimizing path planning when analyzing dynamic electric vehicle path planning that takes battery degradation into account. Additionally, Herle, Channegowda, and Prabhu [8] enhanced the effectiveness of machine learning algorithms in predicting battery state by utilizing deep neural networks to address the issue of limited battery data.

In the study conducted by Schwenk and colleagues [9], they integrated battery aging to the two-way charging optimization process for electric vehicles (EVs). This underlines the significance of considering the effects of battery aging on operational expenses and the futuristic potential of intelligent charging applications for EVs. Furthermore, for the problem of routing electric vehicle batteries, Rodríguez-Esparza and his colleagues [3] proposed a hyper-heuristic algorithm that combines adaptive simulated annealing and reinforcement learning. This method could offer fresh insights into resolving power range restrictions and mitigating travel time and expenses.

A synthesis of the research presented above leads to the following two foci:

1. Quantitative assessment of the environmental impact of new energy vehicle driving behavior.
2. Exploration of the mechanism of the influence of new energy vehicle user behavior on the realization of expected environmental benefits.

For the first focus, it is evident that the driving habits of new energy vehicles have a direct effect on the environment, based on previous research. For the second focus, this study integrates state parameters of new energy vehicles and driver behavior data in simulating urban ecosystems computationally. An LSTM model by utilizing a Bayesian optimization data-driven approach [8] to examine environmental damages brought about by driver behavior patterns. This study also conducts significance tests to understand the long-term impacts in more detail. Furthermore, a significance test is carried out to obtain a more detailed comprehension of the long-term effects.

2 Related Work: Environmental Benefits of New Energy Vehicles

This paper focus on the behavior of new energy vehicle drivers. Battery optimization techniques and power network structures fall outside the scope of our study. As a key measure to reduce pollution emissions, electric vehicles are deemed a promising option [1]. It is crucial to note, however, that the environmental impact of electric vehicles is closely tied to the power supply. The type of energy utilized, such as coal, nuclear, or hydroelectric power, greatly affects the environmental impact of the power supply [3].

According to a life cycle assessment of electric vehicle batteries [13], traveling the same distance with electric vehicles equipped with different battery types results in significantly varying environmental impacts. Lithium-sulfur batteries are deemed the most appropriate option due to their cleaner properties throughout their life cycle. However, a significant number of electric vehicle (EV) users tend to disregard the practical lifespan of their EVs. This negligent behavior results in the abandonment of used batteries in the wild, giving rise to probable environmental hazards [14]. Inadequate disposal of these batteries can inflict a long-term catastrophic impact on the ecosystem, detrimentally affecting plant and animal species, and even posing a threat to human health [15].

Excluding battery recycling and disposal concerns, the scarcity of charging infrastructure is a significant hurdle for the electric vehicle industry. Chen et al. [10] found that the absence of charging stations increases consumer anxiety regarding mileage. Furthermore, studies suggest that unmanaged user behavior could hinder the potential environmental advantages of advanced electric vehicle technology [11].

This paper examines the psychological behavior of electric vehicle users. It discusses the impact of people's perceptions, use paradoxes, range anxiety, and gender attitudes on ownership [17]. The actual use of electric vehicles and consumer purchasing decisions are crucial for achieving their carbon emission benefits [16].

In addition, it is crucial to acknowledge the impact of diverse consumer behaviors and control group vehicle inventories to achieve the emission benefits of electric vehicles [12]. Consequently, assessing the effect of consumer behavior diversity on the realization of environmental advantages of EVs provides valuable insight for developing effective EV purchase incentives in the future [14].

The environmental impact of electric vehicle (EV) battery storage requires thorough investigation. Battery performance declines with time, particularly at high states of charge, which can worsen battery performance and impact its longevity and environmental efficiency [13]. However, little research has examined this matter from the viewpoint of EV owners, creating ambiguity surrounding the application of the term "high state of charge" in practice.

Amidst the increasing global consensus on environmental protection and emissions reduction, this research highlights the significance of driver behavioral improvements, instead of focusing solely on technical optimization, to fully comprehend and actualize the potential of electric vehicles (EVs) [3].

3 Insight of Vehicle Status and Driver Behavior

Current research recognizes common predictions, such as the gradual phasing out of fossil-fueled power plants, lowering emissions during electric vehicle (EV) production, and changes to the automotive energy consumption trend. These forecasts are useful in predicting the future, but they carry the risk of deviating from reality [19]. Our investigation encompasses both renewable and fossil fuel power sources by including electric vehicles and hybrid electric vehicles (HEVs).

To redefine vehicle behavior analysis, a framework that incorporates temporal, spatial, and energy dimensions is developed. The framework aims to provide an objective and comprehensive analysis of vehicle behavior. This data-centric technique aggregates multidimensional data streams which can be mined for insightful characteristics of electric vehicle functionality.

Analyzing multiple types of vehicles, specifically personal electric vehicles, across a variety of temporal and urban classifications reveals intricate driver behavior patterns, charging habits, and vehicle usage [18].

Using this analysis, a detailed investigation into electric vehicle travel data is conducted to comprehend users' driving characteristics thoroughly. This is augmented by ample data from charging stations, facilitating the recognition of electric vehicle (EV) charging dynamics, encompassing time and power metrics that serve as fundamental components for comprehending user charging behavior.

With these fundamental elements in place, the relationship between driving inclinations and charging patterns can be researched. This study on traffic behavior, trip mapping, and charging tendencies offers valuable insights into EV usage from the standpoint of owners.

The process entailed several stages of analysis on vehicle data, initiated with transforming time fields into a timestamp format, which facilitated subsequent aggregation and analysis tasks. Subsequently, a state change analysis was conducted, during which shifts in vehicle states, such as startups and shutdowns, were examined over time in order to comprehend usage patterns. Furthermore, the behavior of the charging mode was also analyzed over time in order to discern the prevalence and effectiveness of different charging strategies, including during driving and parked scenarios.

The operation mode analysis encompassed two principal elements: the distribution of operational modes, such as pure electric versus hybrid usage, and an assessment of driving behaviors corresponding to these modes; and mode switching analysis, which aimed to identify prevalent switching tendencies among vehicles and the conditions prompting such switches.

Outlier detection procedures were implemented in three dimensions: temporal anomalies that flagged irregularities in timestamps, state anomalies detecting unconventional vehicle or charging status changes, such as immediate shutdown post-startup, and mode anomalies which highlighted inconsistencies in operation mode shifts, for instance, transitioning to fuel mode during a charging session. Moreover, feature-based analyses of driver behavior were conducted, with driving and charging modes serving as bases. These included assessments of whether drivers who frequently opt for pure electric mode exhibit more eco-friendly practices and of the circumstances under which drivers choose to utilize driving as a means to charge the vehicle.

Correlative insights between the state of charge (SOC) of the electric vehicle over time and several pivotal indicators were visually presented through correlation analyses, encapsulated in two diagrams. These diagrams further enrich the understanding of the interplay between SOC management and overall vehicle performance (Fig. 1).

There is a positive correlation between the state of charge (SOC) and the cumulative mileage (sum of mileage) of the vehicle, showed in Fig. 2. Suggesting that higher battery SOC could lead to greater mileage.

Additionally, there is a positive correlation between SOC and the sum of voltage, indicating a relationship between battery SOC and voltage.

A negative correlation exists between SOC and battery current. This implies that when the battery's state of charge is high, the battery current is low. And, a negative correlation exists between SOC and vehicle power. This means that when the battery's state of charge is high, the vehicle power decreases.

A negative correlation exists between distance and SOC, suggesting that the battery's charge decreases as the vehicle travels a greater distance, showed in Fig. 2. Besides, a

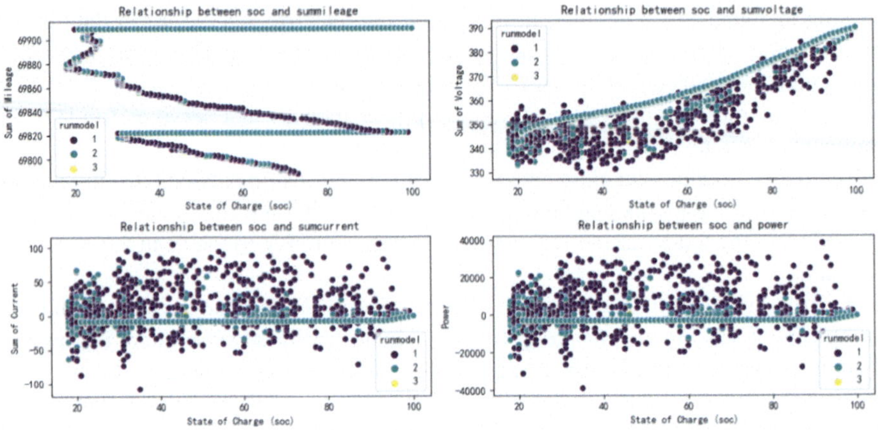

Fig. 1. Relationship between soc and sum-mileage, sum-voltage, sum-current, power

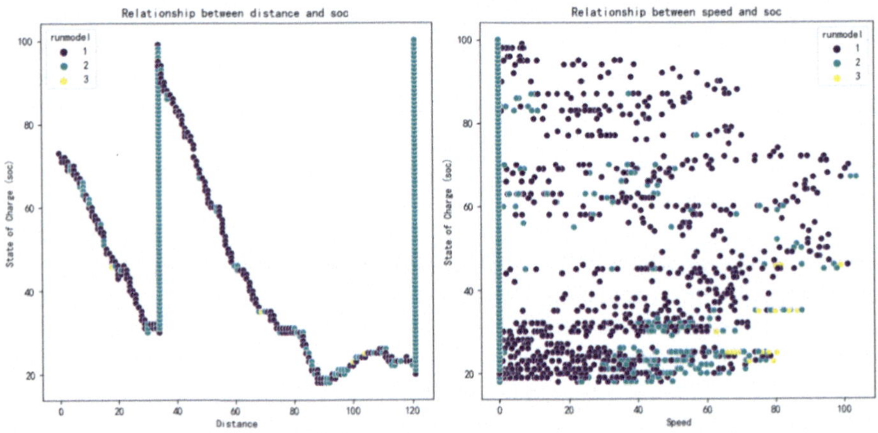

Fig. 2. The relationship between soc and distance, soc and speed

negative correlation exists between speed and SOC, indicating that the battery's charge decreases as the vehicle travels faster.

4 Urban Ecosystem Modeling

An all-encompassing assessment model has been established to evaluate the ecological effects of land degradation, pollutant concentrations, climate change, forest rehabilitation, waste management, biodiversity, levels of industrialization, traffic expansion, energy usage, new energy transportation, and the chemical toxicity of lithium-sulfur batteries.

The model aims to examine the use of lithium-sulfur batteries, their chemical toxicity, and the environmental impact of carbon emissions resulting from the utilization of electric energy in new energy cars. Collect and process data using appropriate preprocessing and feature engineering techniques. To incorporate new energy vehicles into this ecosystem model, the electrical consumption by these vehicles was converted to carbon emissions, following international standards. Harmful substances produced in battery production were also quantified, using international standards.

An LSTM was trained using a Bayesian parameter optimizer, and this LSTM was used to extend all data in the time dimension. All weight factors in this model were acquired through the information Entropy Weighting Method (EWM). The model illustrates the relationship between new energy vehicles and the urban environment through the use of simulation modeling. These indicators and interactions are expressed through the following mathematical equations.

4.1 Data Extending

The following graphs show how the LSTM works with our data, including the training process in Fig. 3. And the fit performance of the LSTM in Fig. 4.

The model has shown a good fit with the historical data, indicating that its predictions for future data points are likely to be reliable. This strong match between the model's output and the observed historical data provides a solid basis for making forecasts.

By using the future data generated by the model, combined with relevant indicators, we can simulate how the ecosystem might evolve over time. Both long-term trends and short-term variations, as predicted by the model, allow us to create a better picture of the potential changes in the ecosystem. This approach helps us to better understand the possible directions in which the ecosystem could develop and the various factors that might influence these changes. Through such simulations, we gain valuable insights into the future state of the ecosystem, enabling more informed decision-making and planning.

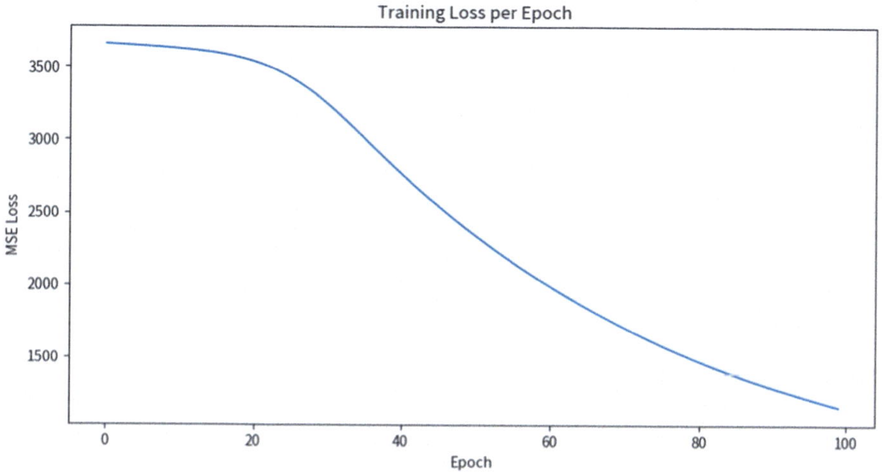

Fig. 3. Decline curve of training loss

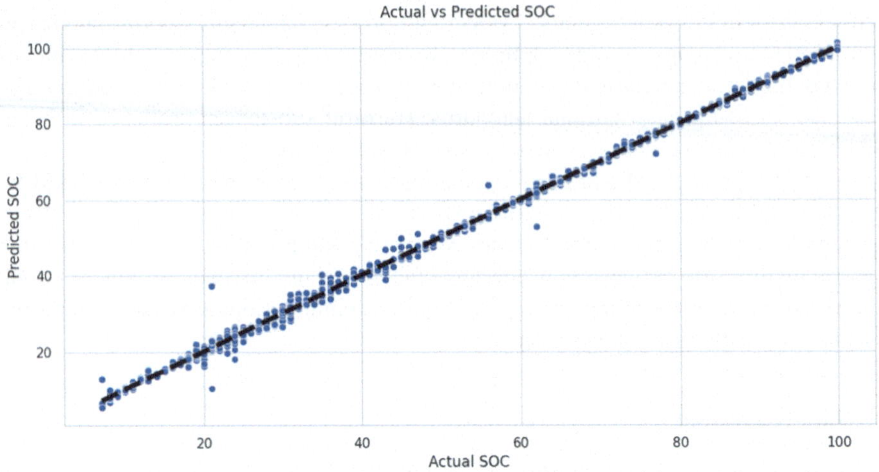

Fig. 4. Test goodness of fit

Table 1 indicates the model exhibits comparable goodness of fit during training and testing. However, the mean squared error (MSE) during training is approximately half that observed during testing. This may suggest that the model is exhibiting signs of overfitting.

Table 1. Evaluation of LSTM with Bayes-Optimizer

Metric		Value
R2	Train	0.999274
	Test	0.998679
MSE	Train	0.497713
	Test	0.938392

4.2 Simulating

The evaluation indicators are summarized [21, 22]. All the ϵ symbols below denote random error terms.

Land Degradation (LD)

$$LD = l_1 \cdot Pop + l_2 \cdot T_r + l_3 \cdot \alpha + l_4 \cdot A + \varepsilon_{LD} \quad (1)$$

(Pop) Population density.
(T_r) Rate of change of regional temperature.

(α) Parameter of land use pattern, reflecting the type and intensity of land use.
(A) The size of the area that may influence land degradation.

Pollutant Concentration (PC)

$$PC = g_1 \cdot C_{industrial} + g_2 \cdot C_{traffic} + g_3 \cdot C_{waste} + g_4 \cdot RAD + \varepsilon_{PC} \tag{2}$$

($C_{industrial}$) Industrial pollutant concentration.
($C_{traffic}$) Concentration of pollutants caused by traffic.
(C_{waste}) Concentration of pollutants caused by waste.
(RAD) Residential Area Disturbance Degree (RAD): The impact of residential activities on the environment.

Climate Change Impact (CCI)

$$CCI = c_1 \cdot \Delta Temperature + c_2 \cdot \Delta Precipitation + \varepsilon_{CCI} \tag{3}$$

($\Delta Temperature$) Temperature difference.
($\Delta Precipitation$) Precipitation difference.

Forest Restoration (FR)

$$FR = d_1 \cdot \text{VFC} + d_2 \cdot \text{DI} + d_3 \cdot R + \varepsilon_{FR} \tag{4}$$

(VFC) Vegetation cover.
(DI) Forest Destruction Index.
(R) Forest recovery rate.

Waste Treatment Rate (WTR)

$$WTR = e_1 \cdot C_{\text{waste}} + e_2 \cdot P + \varepsilon_{WTR} \tag{5}$$

(C_{waste}) Concentration of waste generated.
(P) Waste treatment policy and treatment efficiency.

Lithium Sulfur Battery Chemical Toxicity (LSBCT)

$$\begin{aligned} LSBCT = {} & g_1 \cdot Charging\,Behavior + g_2 \cdot Temperature \\ & + g_3 \cdot Driving\,Speed + \varepsilon_{LSCBT} \end{aligned} \tag{6}$$

In the Lithium Sulfur Battery Chemical Toxicity (LSBCT) equation, the function (p) establishes a relationship between charging behavior, temperature, and driving speed. This function serves as a model for evaluating the potential environmental impacts resulting from improper use of lithium-sulfur batteries.

Ultimately, these impacts are assigned weight values and added together to calculate the NEV Efficiency.

NEV Efficiency

$$E = w_1 \cdot LD + w_2 \cdot PL + w_3 \cdot CCI + w_4 \cdot FR + w_5 \cdot WTR + w_6 \cdot NEVI + w_7 \cdot LSBCT + \varepsilon_E \quad (7)$$

The above model highlights that overcharging and discharging, as well as the habit of frequent recharging only when the battery is very low, significantly increase the chemical toxicity emissions from lithium-sulfur batteries.

Furthermore, operating a lithium-sulfur battery at high speeds can cause overheating and a higher risk of harmful chemical emissions.

Parameters Setting

$$NEV\ Population = 31.6\% \quad (8)$$

The heightened chemical toxicity of lithium-sulfur batteries not only harms battery performance but also negatively impacts the environment, increasing the urgency for implementing technology and modifying behaviors to mitigate traffic-related pollution.

Simulating Results

After simulating through these indicators, the following results were obtained. Results of Environmental Impact is showed in Table 2, while results of NEV Efficiency is showed in Table 3.

Table 2. Results of Environmental Impact

Indicator	Before simulation	After simulation
Land Degradation (LD)	3.22	3.53
Pollutant Concentration (PC)—SO_2	0.04 mg/m^3	0.07 mg/m^3
Climate Change Impact (CCI)	0.21 °C	0.32 °C
Forest Restoration (FR)	0.32%	0.26%
Waste Treatment Rate (WTR)	20.3%	15.5%
Lithium Sulfur Battery Chemical Toxicity (LSBCT)	4	4

Table 3. Results of NEV Efficiency

Indicator	Values	Standardization
NEV Growth (Δ NEV Population)	0.001321	1
Δ NEV Efficiency	−0.004252	−3.21877

5 Discussion

During our analysis and mining of locomotive state and driver behavior data, it was observed that electric vehicles exhibited phenomena such as high-speed driving and battery overloading. Based on the correlation between the state of charge (SOC) and related indexes, an increase in driving speed leads to a reduction in battery charge, an increase in battery power, and subsequently results in overheating of the batteries, causing melting and production of toxic substances. These actions shorten battery life and adversely affect the environment. Charging the battery only until it reaches its low limit or constantly fully charging it can harm the battery's performance and reduce its lifespan.

NEV Growth, which is 0.001321, indicates that an increase in the per-unit popularity of new energy vehicles will have a slightly positive impact on the environment, specifically increasing landscape diversity and reducing disturbance in residential areas.

However, enhancements in NEV efficiency do not necessarily result in expected environmental benefits.

According to NEV Efficiency which is −0.004246, technological advancements may not offset the negative environmental effects of inappropriate vehicle use habits. Thus, improved efficiency does not inevitably lead to environmental benefits.

The driving habits of electric and hybrid vehicle operators suggest that they tend to overcharge and overuse electricity, resulting in damage to the battery and the surrounding ecosystem. By applying long short-term memory (LSTM) to the timeseries data of automobiles and extending it to the end of their lifecycle, this information is integrated into our ecological evaluation model calculations. The outcome of this process revealed that the implementation of new energy vehicles does not significantly enhance the environment, thus indicating that these vehicles do not fulfill the anticipated goals of effective energy conservation and emissions reduction. This study suggests that new energy vehicles do not meet the expectations of effectively saving energy and reducing emissions.

The research primarily focuses on vehicle owners' behavioral patterns to understand this phenomenon. Furthermore, current research has not determined the optimal use of lithium-sulfur batteries, which are commonly used in new energy vehicles. This is a crucial area for further investigation. In addition, future policies must be implemented to improve hardware and software related to monitoring driver speed. Finally, a significant advancement is made in computationally analyzing the interaction between electric vehicles and the environment with urban ecosystem modeling. These findings hold substantial importance.

6 Conclusion

In this study, an analysis is conducted on the alteration of parameters concerning two emerging energy vehicles, namely electric and hybrid vehicles, in conjunction with driver behavior. The research shows that certain driver behavior such as overcharging, discharging, and excessive speed have the potential to be detrimental to the battery, thereby posing a risk to the environment. Based on our findings, a simple yet effective computational simulation model is proposed for urban ecosystems. It covers data generation, evaluation, and analysis. The results of our simulations confirm that poor driving habits have adverse effects on urban ecology. The work encourages relevant authorities to improve driving behaviors. Additionally, the urban ecosystem simulation model offers an excellent testing environment for simulation experiments.

References

1. Hsu, T.: On the sustainability of electrical vehicles (2013). ArXiv:abs/1311.6015
2. Vitta, S.: Electric cars—assessment of 'green' nature vis-à-vis conventional fuel driven cars. Sustain. Mater. Technol. (2021)
3. Rodr'iguez-Esparza, E., Masegosa, A.D., Oliva, D., Onieva, E.: A new hyper-heuristic based on adaptive simulated annealing and reinforcement learning for the capacitated electric vehicle routing problem (2022). ArXiv:abs/2206.03185
4. Mousaei, A.: Improving Energy Management of Hybrid Electric Vehicles by Considering Battery Electric-Thermal Model (2023). arXiv:2302.13157
5. Sitapure, N., Kulkarni, A.: Exploring Different Time-series-Transformer (TST) architectures: a case study in battery life prediction for Electric Vehicles (EVs) (2023). Arxiv:abs/2308.03260
6. Cui, X., Ramyar, A., Siegel, J.B., Mohtat, P., Stefanopoulou, A.G., Avestruz, A.: Comparing power processing system approaches in second-use battery energy buffering for electric vehicle charging. J. Energy Storage (2021)
7. Herle, A., Channegowda, J., Prabhu, D.: Overcoming limited battery data challenges: a coupled neural network approach. Int. J. Energy Res. **45**, 20474–20482 (2021)
8. Gebbran, D., Rich, J., Dragičević, T.: Electrical vehicle fleet routing accounting for dynamic battery degradation (2022). ArXiv:abs/2203.04642
9. Meisenbacher, S., Schwenk, K., Galenzowski, J., Waczowicz, S., Mikut, R., Hagenmeyer, V.: Smart charging of electric vehicles with cloud-based optimization and a lightweight user interface: a real-world application in the energy Lab 2.0: poster. In: Proceedings of the Twelfth ACM International Conference on Future Energy Systems (2021)
10. Chen, S., Olivioufski, E., Dods, B.G., Hamdi, M.: Multi-period coordinated management of electric vehicles in zonal power markets: a convex relaxation approach. Optim. Control (2017)
11. Bortoli, A.D.: Environmental performance of shared micromobility and personal alternatives using integrated modal LCA. Transp. Res. Part D: Transp. Environ. (2021)
12. de Bortoli, A., Agez, M.: Environmentally-extended input-output analyses efficiently sketch large-scale environmental transition plans: illustration by Canada's road industry. J. Clean. Prod. (2023)
13. Gu, Y., Liu, M., Naoum-Sawaya, J., Crisostomi, E., Russo, G., Shorten, R.: Pedestrian-aware engine management strategies for plug-in hybrid electric vehicles (2016). ArXiv:1612.07568.pdf
14. Juarez, J., Flores, W.P., Lu, Z., Hattori, M., Hernandez, M., Larios-Ramirez, S., Woo, J.: Consumer's Behavior Analysis of Electric Vehicle using Cloud Computing in the State of New York (2023). ArXiv:abs/2306.01888
15. Meyer, D., Wang, J.: Integrating ultra-fast charging stations within the power grids of smart cities: a review. Optim. Control (2018)
16. Sivagnanam, A., Ayman, A., Wilbur, M., Pugliese, P., Dubey, A., Laszka, A.: Minimizing energy use of mixed-fleet public transit for fixed-route service (2020). ArXiv:abs/2004.05146

17. Wang, G., Chen, X., Zhang, F., Wang, Y., Zhang, D.: Experience: understanding long-term evolving patterns of shared electric vehicle networks. In: The 25th Annual International Conference on Mobile Computing and Networking (2018)
18. Baghali, S., Guo, Z., Hasan, S.: Investigating the spatiotemporal charging demand and travel behavior of electric vehicles using GPS data: a machine learning approach (2022). arXiv:2203.00135
19. Ledesma-Alonso, R., Becerra-Nunez, G.: Electric or gasoline: a simple model to decide when buying a new vehicle (2023)
20. Zhang, K., Lygeros, J.: Routing and charging game in ride-hailing service with electric vehicles (2023). arXiv:2309.05120
21. Balzan, M.V.: Assessing ecosystem services for evidence-based nature-based solutions (2021)
22. Boeing, G.: Measuring the complexity of urban form and design. Urban Des. Int. **23**, 281–292 (2017)

DFT Study on Green and Environmentally Friendly Production of 1,3-PDO Through Glycerol Catalytic Hydrogenation on the Pt/WO_3(1 0 0) Surface

Yakun Qu(✉), Xiaoguang Zhao, Lixin Wang, and Chao Jin

Research Institute of Petroleum Processing, Sinopec, Beijing, China
{quyakun.ripp,zhaoxg.ripp,wanglx.ripp,jinchao.ripp}@sinopec.com

Abstract. The quick development of biodiesel industries has created a huge surplus of glycerol. Glycerol is a renewable Bioenergy material, which is widely used in the chemical industry. For environmental protection, its environmental performance is excellent. Under this situation, considerable attention has been focused on finding reasonable and effective ways to convert glycerol into chemicals. Currently, the most promising method for converting glycerol is the hydrogenolysis of glycerol which produces 1,3-propanediol. In the process of producing 1,3-PDO by traditional chemical industrial synthesis, the intermediate acrolein itself is highly toxic and harmful to human health. The preparation of 1,3-propanediol through glycerol conversion is an environmentally friendly and cost-effective route, with very broad applications and great development prospects. However, one of the biggest challenges that this method needs to overcome is to develop a catalyst with high yield and low cost, so that the main product of glycerol hydrogenation reaction is 1,3-PDO and relatively high yield can be achieved. This paper studies the role of noble metal Pt in catalyzing hydrogenation and hydrolysis of glycerol to produce 1,3-PDO in the Pt/WO_3 system using DFT calculations. The spatial distributions of electron densities on the catalyst surface and Mulliken population analysis indicate that Pt can enhance the acid strength of Brønsted acid sites at the ortho position on the WO_3 surface, lower the activation energy for active-phase reactions, reduce the activation energy for glycerol hydrogenation, and heighten the selectivity to hydrogenate the secondary hydroxyl group. As per the frontier molecular orbital theory, there is a high match of orbitals between the Brønsted acid sites in the vicinity of Pt on the WO_3 surface and the secondary hydroxyl group of the glycerol molecule, and a small energy level difference between the frontier orbitals, thus lowering the activation energy to protonate the intermediate secondary hydroxyl group of glycerol and increasing the selectivity to 1, 3-PDO. In order to achieve high 1,3-PDO selectivity and yields, an appropriate combination of Pt loading and Brønsted acid sites should be adopted in the design of catalyst. Furthermore, a suitable number of Brønsted acid sites in the vicinity of Pt should be added so as to selectively activate the intermediate secondary hydroxyl group of glycerol and increase the yields and selectivity of 1,3-PDO, which not only solves the problem of glycerol excess, but also provides a green way to produce 1,3-propanediol, providing basic data for the industrial application of catalysts.

G. Huang et al. (Eds.): ICESD 2024, ESE, pp. 156–173, 2025.
https://doi.org/10.1007/978-3-031-88683-6_12

Keywords: DFT · Glycerol · 1,3-propanediol · Biofuels · Waste valorization · Renewable sources of energy

1 Introduction

The quick development of biodiesel industries has created a huge surplus of glycerol. Glycerol is a renewable Bio-energy material, which is widely used in the chemical industry. For environmental protection, its environmental performance is excellent. In view of this, considerable attention has been focused on finding reasonable and effective ways to convert glycerol into chemicals. Currently, the most promising method for converting glycerol is the hydrogenolysis of glycerol which produces 1,3-propanediol, and the key work of glycerol hydrogenolysis is the research on its catalytic mechanisms and catalyst-related design [1].

Biodiesel is a renewable and environmentally friendly alternative to non-renewable fossil fuels. In order to produce biodiesel, transesterification of vegetable oils and animal fats can be carried out, and in the transesterification process glycerol as the main by-product produced accounts for about 10% of the total biodiesel produced [2]. With the continuously expanding production and using of biomass, glycerol has become more accessible as a renewable energy source. Researchers are now preferring to developing new technologies for converting glycerol into value-added chemicals.

1,3-Propanediol (1,3-PDO) is an important commercial chemical widely used in the pharmaceutical, food and textile industries [3]. At present, 1,3-PDO is mainly used as a monomer in the production of polymers, such as polytrimethylene terephthalate and polyurethane, etc. [4]. The market price of 1,3-PDO is about $4000 per ton [5–12]. Now its using as a polymer is wide spreading, and leads to large market growth, 1,3-PDO may make the transition from a traditional specialty chemical to a commodity chemical, with a significant reduction in production costs.

The traditional chemical synthesis of 1,3-PDO by glycerol involves two methods, one is Degussa and the other is Shell processes. The Degussa process was proposed by DuPont in 1998 and subsequent production was discontinued [13]. Degussa uses acrolein as the raw material. First, acrolein is hydrated in the aqueous solution at a temperature of 50–70 °C to produce 3-HPA, and 3-HPA is hydrogenated on a nickel catalyst at a pressure of 135 bar and a temperature of 75–145 °C [14, 15]. The Shell process, on the other hand, uses ethylene oxide and synthesis gas to produce 3-HPA through the hydroformylation reaction, and then in the presence of a ruthenium, rhodium or cobalt-based catalyst, 3-HPA is heated at 100 atm and 100–150 °C and hydrogenated to 1,3-PDO [14, 16]. However, the production of 1, 3-PDO by hydration of acrolein followed by hydrogenation is highly toxic and danger to human health, while the production of 1,3-PDO by the ethylene oxide carbonization method involves high equipment costs, technological challenge, and complex catalyst syntheses.

The production of 1,3-PDO from glycerol is environmentally friendly, relatively low-cost, and has good prospects in the development. Nevertheless, one of the biggest challenges for realizing this method is to develop a high-yield and low-cost catalyst, so that 1,3-PDO instead of 1,2-PDO is formed in glycerol hydrogenation with relatively high yields.

In the reaction pathway for hydrogenation of glycerol to produce propylene glycol, the acidity and basicity of the reaction system has greatly affected the course of the reaction. An acid or neutralization reaction takes place in three steps according to the carbocation mechanism [17–19] The first step is protonation of a hydroxyl group of glycerol, a dehydration reaction is the second step; the third step is hydrogenation in the metal's active sites and hydride-ion transfer to generate propylene glycol. If the secondary hydroxyl group first gets an H^+, the propylene glycol produced would be 1,3-PDO, and if the primary hydroxyl group first gets an H^+, the propylene glycol produced would be 1,2-PDO.

The technical challenge for optimizing the hydrogenolysis of glycerol to 1,3-PDO lies in selectively controlling the location of hydroxyl groups in glycerol protonation [20]. Due to the presence of two primary hydroxyl groups, the secondary hydroxyl group has lower reactivity in space, making it less accessible to the active sites of the catalyst. For this reason, it is kinetically and thermodynamically disadvantageous to remove secondary hydroxyl groups [21, 22].

In order to efficiently catalyze the hydrogenation and hydrolysis of glycerol to produce 1,3-PDO, researchers have explored a heterogeneous group of bifunctional catalysts that contain both metal and acid catalytic functions [23, 24]. To achieve higher selectivity to 1,3-PDO, one needs to find suitable combinations of metal sites and Brønsted acid sites so as to selectively activate secondary hydroxyl groups of glycerol. The effect of non-noble metals as active hydrogenation sites on the hydrogenolysis of glycerol to synthesize 1,3-PDO is not remarkable [18, 25–32], whilst noble metals act well as highly active hydrogenation sites, so Pt, Ir, Au, Rh and other noble metals are applied in this reaction for preparing catalysts [22, 33–40]. Among the noble metallic elements, Pt is the most widely used [28, 41–75]. Metal sites activate hydrogen molecules, and acid sites play a key role in activating glycerol. Metallic oxides have affinities for trialcohol hydroxyls [21, 24]. Among the Brønsted acids, WO_x oxides demonstrate to exert the best effect on the selective conversion of glycerol into 1,3-PDO [22]. At the same time, the acid amount and strength of Brønsted acid sites are vital for attaining higher glycerol adsorption and conversion.

In this study the role of noble metal Pt in catalyzing the hydrolysis and hydrogenation of glycerol to produce 1,3-PDO in the Pt/WO_3 system is examined through using DFT methods, which provides theoretical guidance for improving catalytic activity and selectivity, and points out the direction for the design and development of new catalysts from the perspective of cost reduction for noble-metal-based catalysts, which not only solves the problem of glycerol excess, but also provides a green way to produce 1,3-propanediol, providing basic data for the industrial application of catalysts.

2 Method

All calculations were carried out with the program package in Materials Studio 2020. In $DMol^3$, the physical wave functions are expanded in terms of accurate numerical basis sets. We used the double-numeric quality basis set with polarization functions. The generalized gradient approximation (GGA) with the revised Perdew–Burke–Ernzerhof functional (RPBE) was utilized. The RPBE functional usually gives absolute errors in adsorption energy of about 0.2 eV. A Fermi smearing of 0.0005 a.u. (0.0136 eV). For numerical integration, the medium quality mesh was used, i.e. the convergence criteria for structure optimization and energy calculation were set to ① SCF tolerance of 1×10^{-5} a.u./ atom, ② energy tolerance of 2.0×10^{-5} a.u./atom, ③ maximum force tolerance of 4.0×10^{-3} a.u./ Å and ④ maximum displacement tolerance of 5.0×10^{-3} Å. The complete linear synchronous transit and quadratic synchronous transit (LST/QST) method was used to locate the transition states [1–4].

3 Results and Discussion

3.1 The Catalyst Model

All the work in this paper is carried out on the surface of hexagonal-WO_3(1 0 0), and the WO_3(1 0 0) cluster model established is as shown in Fig. 1. On the WO_3(1 0 0) surface, the O atoms have an unsaturated coordination, so they are coordinated in 1, and the W atom possesses saturated coordination, so it is coordinated in 6.

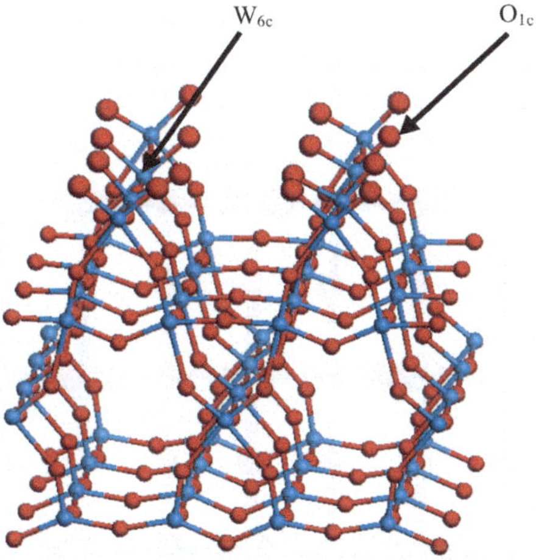

Fig. 1. Hexagonal-WO_3(1 0 0) cluster model

In order to determine the loading form of Pt atoms on the WO_3(1 0 0) surface, the single point energies of Pt atoms at four different loading positions on the WO_3(1 0 0) surface are calculated, as shown in Fig. 2. Generally speaking, a structure with less single point energy would be more stable. The structure in Fig. 2a has the lowest energy level and is relatively stable, and Pt is bonded to two unsaturated coordinated O atoms on the WO_3(1 0 0) surface. In agreement with the structural model in Fig. 2b, Pt is bonded to four unsaturated coordinated O atoms on the WO_3(1 0 0) surface, and its energy is 15.587 kcal/mol higher than that of the structural model shown in Fig. 2a. The structural models are shown in Fig. 2c, d which have higher energy levels and are more unstable relatively. Therefore, the structural model displayed in Fig. 2a is taken as the H_2 dissociation calculation model.

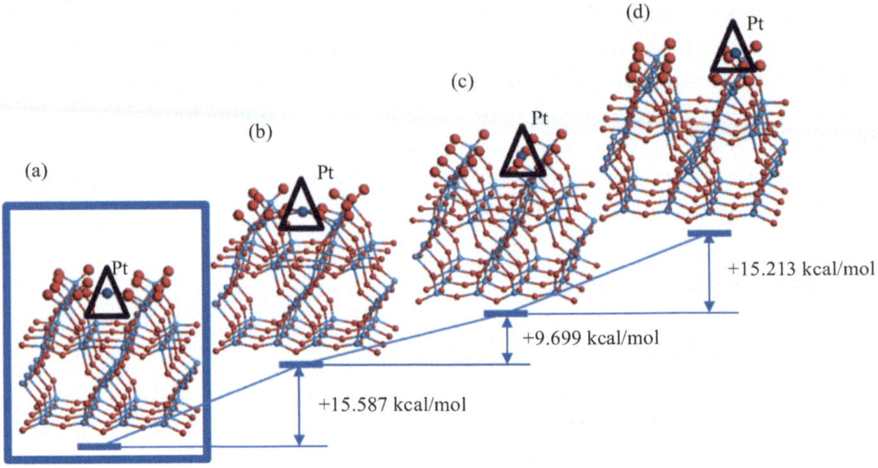

Fig. 2. Pt/WO_3(1 0 0) cluster model

Figure 3 presents the top view of the structural model in Fig. 2a, where the O atom is at the ortho position of the Pt atom as they occupy positions next to each other, and is at the meta position of the Pt atom as they are at the 1 and 3 positions.

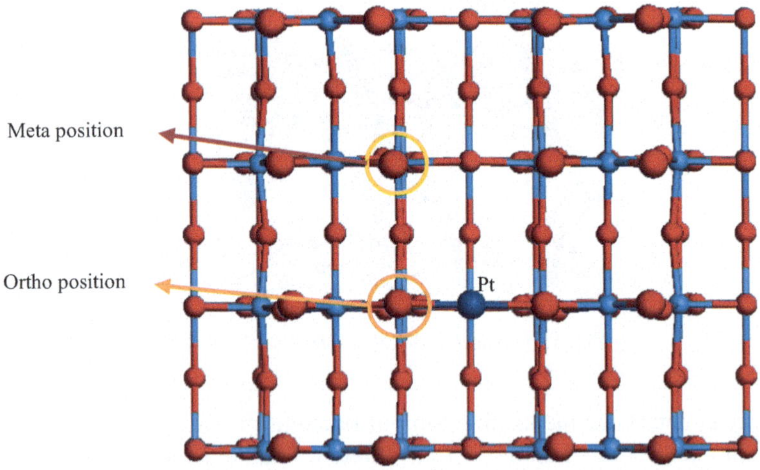

Fig. 3. The top view of a stable Pt/WO_3(1 0 0) cluster model

3.2 Active-Phase Formation

The Brønsted acid sites in WO_3 play a key role in the selective hydrogenolysis of glycerol. The number of active sites must be proportional to the number of active metal sites to maximize the selectivity to 1,3-PDO. Acid strength is another important factor, because strong acid sites can result in intense reactions to alcohol, which decrease product selectivity, whilst weak acid sites would lower the conversion rate and weaken catalytic activity. In order to avoid the production of undesirable products in the conversion of glycerol and to orient the reaction path to increase the selectivity to 1,3-PDO, it is fundamental to understand the specific active sites of the catalysts involved in each reaction step. Hence, the formation of Brønsted acid sites is first explored.

H_2 dissociation. In the formation of Brønsted acid sites on the WO_3 surface, the first step is the dissociation of H_2. To begin with, the dissociation of H_2 on the WO_3(1 0 0) surface loaded with and without Pt are calculated, which is as shown in Fig. 4. When H_2 dissociates on the WO_3(1 0 0) surface without Pt, the structure of the reactant is R1, the structure of the product is P1, and the activation energy of the reaction is 79.694 kcal/mol. When H_2 dissociates on the Pt-loaded WO_3(1 0 0) surface, the structure of the reactant is R2, the structure of the product is P2, and the activation energy of the reaction is 7.058 kcal/mol. The comparison shows that Pt modifications can dramatically decrease the dissociation of H_2 on the WO_3 surface. In other words, the activation energy for the first-step reaction in the formation of Brønsted acid sites has enhanced catalytic activity.

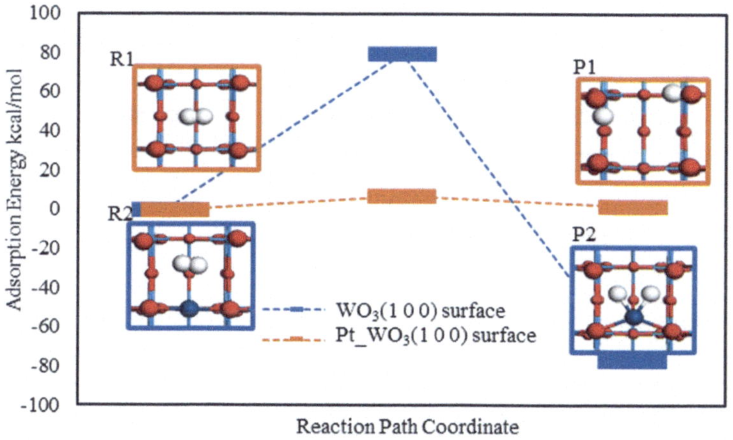

Fig. 4. The dissociation of H_2 on the WO_3(1 0 0) surface loaded with and without Pt

Hydrogen spillover and formation of Brønsted acid sites. After H atoms are generated on the Pt site, it can spill over to the WO_3 active site, forming different Brønsted acid sites and therefore having different catalytic activities. When spilling over to the O atoms on the WO_3 surface next to the Pt atoms, the Brønsted acid site at the ortho position of Pt will be formed, and when spilling over to the O atoms that occupy the 1 and 3 positions with Pt, the Brønsted acid site at the meta position of Pt will be formed. The processes of forming different Brønsted acid sites when H spills over to different positions are calculated, as presented in Fig. 5. The spillover of H atoms is exothermic, so the whole process is thermodynamically supported. In addition, the total heat release (69.556 kcal/mol) in the formation of the Brønsted acid site at the meta position is greater than that (40.907 kcal/mol) at the ortho position, which indicates that the Brønsted acid site at the meta position is more stable. The structure of the Brønsted acid site at the meta position of Pt is P6, while the structure of the Brønsted acid site at the ortho position of Pt is P5. From a kinetic point of view, the activation energy required to form the Brønsted acid site at the ortho position of Pt is 34.127 kcal/mol, which is greater than that at the meta position of Pt, only 7.363 kcal/mol, suggesting that the spillover of H atoms makes it easier to form the Brønsted acid site at the meta position of Pt.

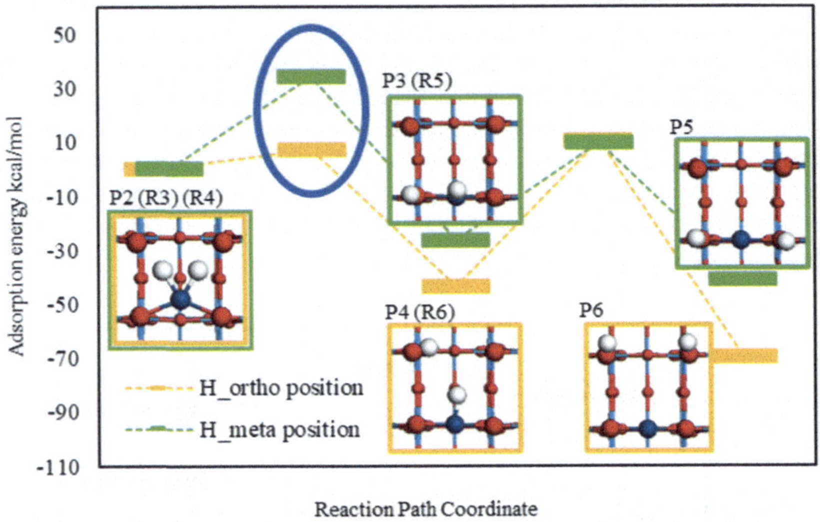

Fig. 5. The formation of Brønsted acid sites with H spillover to different positions

When the feedstock H_2 accounts for more, H atoms may first spill over to the O atom at the meta position of Pt. The structure of the Brønsted acid site is denoted by P7 in Fig. 6, and the structure before H atom spillover is denoted by R7 in Fig. 6. At this time, the activation energy required for H spillover to the ortho position of Pt is only 6.398 kcal/mol, which is significantly lower than the activation energy required for H spillover to form the Brønsted acid site at the ortho position of Pt on the smooth WO_3 surface. In this case, it is easier to form the Brønsted acid site at the ortho position of Pt.

3.3 The Acid Strength of Brønsted Acid Sites in Different Microchemical Environments

The strength of Brønsted acids is generally measured by its ability to donate H protons. On the WO_3 surface, the more positive charges H carries, the greater the degree of O–H bond polarization and the easier it is to release H protons. The acid strength is studied by calculating the Mulliken charge populations of the Brønsted acid sites in different microchemical environments, as demonstrated in Fig. 7. In the Brønsted acid site at the ortho position of Pt (Fig. 7a), the H atom has a positive charge of +0.345e at most. The distance between the Brønsted acid site and the Pt atom shown in Fig. 7b is slightly

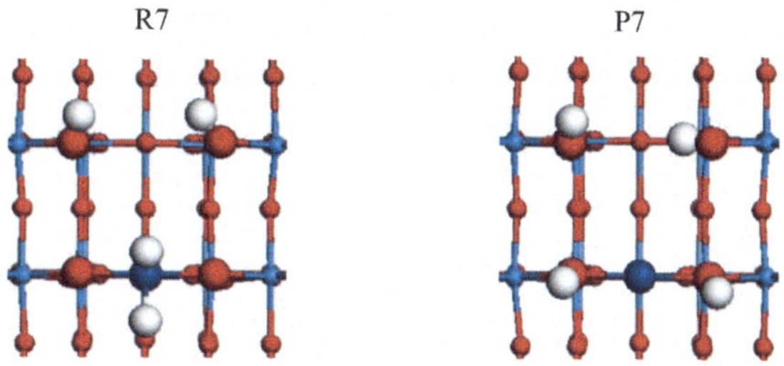

Fig. 6. H atom spillover when the ortho position of Pt is covered by H

longer than that in Fig. 7a, and the H atom in the Brønsted acid site shown in Fig. 7b has a smaller positive charge of +0.325e, indicates that the acid strength also will be reduced. The H atom in the Brønsted acid site at the meta position of Pt (Fig. 7c) has a positive charge of +0.320e, which is lower than that at the ortho position of Pt (Fig. 7b), and the acid strength is also weaker. The Brønsted acid site on the WO_3 surface not loaded by Pt is shown in Fig. 7d, and the H atom carries a positive charge of +0.310e. The distance between the Pt atom and the Brønsted acid site (Fig. 7e) is slightly longer than the distance between the Pt atom and the Brønsted acid site at the meta position of Pt (Fig. 7c), and the H atom carries a smaller positive charge of +0.307e, which is smaller than the H atom in the Brønsted acid site on the WO_3 surface not loaded by Pt (Fig. 7d), and the acid strength is weaker. In a nutshell, Pt atoms can increase the acid strength of the adjacent Brønsted acid sites; however, as the distance between the Brønsted acid site and the Pt atom gets longer, the acid strength will be decreased; and when the distance between the Brønsted acid and the Pt atom is greater than 6.269Å, while the acid strength will decrease slightly.

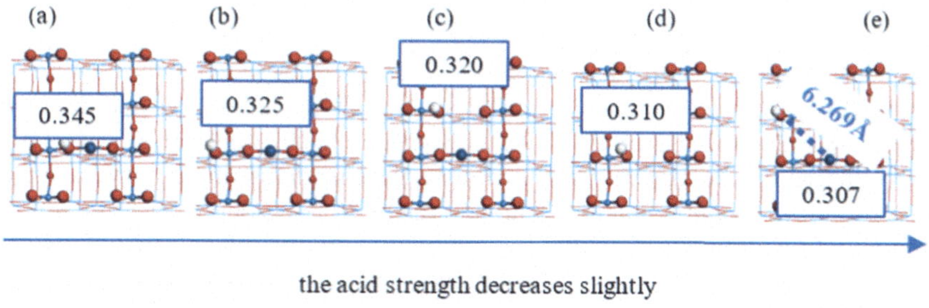

Fig. 7. Mulliken charge populations of H atoms in Brønsted acid sites in different microchemical environments

The same conclusion can be drawn through analyzing the electron densities of Brønsted acid sites in different microchemical environments, as shown in Fig. 8. On the equipotential surface where the electrostatic potential is zero, according to the spatial distributions of the electron densities of Brønsted acid sites in different microchemical environments, the Brønsted acid site at the ortho position of Pt has low electron density (dark blue), and the Brønsted acid site without Pt loading has relatively high electron density (blue). In general, the Brønsted acid site on the WO_3 surface has the lowest electron density in the entire cluster and can interact electrostatically with other electron-rich molecules.

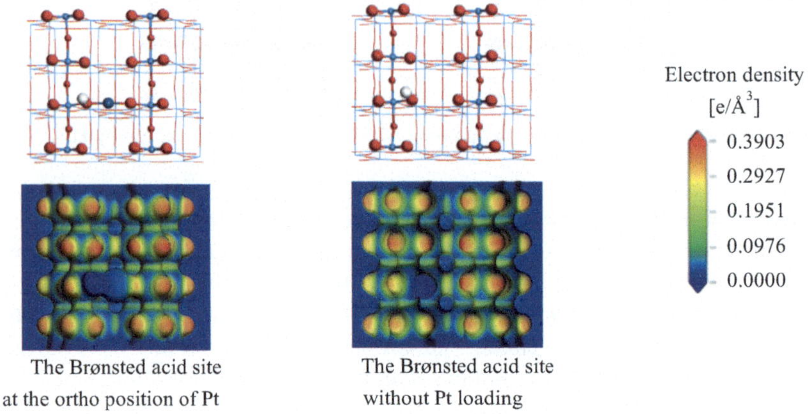

Fig. 8. The effect of Pt loading on the electron densities of Brønsted acid sites on the WO_3 surface

3.4 Hydrogenation and Hydrolysis in Different Acid Sites

Glycerol is a feedstock to produce 1,3-PDO, and its properties directly affect the yield of 1,3-PDO. Therefore, the chemical properties of glycerol are first studied. As per Figs. 9 and 10, in the glycerol molecule, the secondary C2—OH bond order is smaller than the primary C1(3)—OH bond order, indicating that the secondary C2—OH bond is easier to break. Meanwhile, the Mulliken charge population analysis shows that the O atom in the secondary hydroxyl group has a negative charge of −0.603e while the O atom in the primary hydroxyl group has a negative charge of −0.571e, signifying that the O atom in the secondary hydroxyl group will have a stronger electrostatic interaction with the positively charged Brønsted acid site. Obviously, the spatial distributions of the electron densities of glycerol exhibit that the O atom in the hydroxyl group is the close target for the H atom in the Brønsted acid site. However, it is not difficult to conclude from the molecular structure of glycerol that the secondary hydroxyl groups have greater steric hindrance.

Based on the above research on the properties of the Brønsted acid sites and the glycerol molecule in the Pt/WO_3 system, the activation energy and enthalpy change for the hydrogenation and hydrolysis at different acid sites are calculated, as shown in Table 1. At the Brønsted acid site in the ortho position of Pt, Reaction (8) is a protonation reaction of the secondary hydroxyl group in the middle of glycerol ($E_a = 36.322$ kcal/mol), and Reaction (9) is the corresponding hydride-ion transfer ($E_a = 22.407$ kcal/mol); the energy barrier for protonation is higher than that for hydride-ion transfer, which suggests that the protonation reaction is the rate-determining step in the hydrogenation and hydrolysis of glycerol to 1,3-PDO. In the same way, a similar conclusion can be drawn

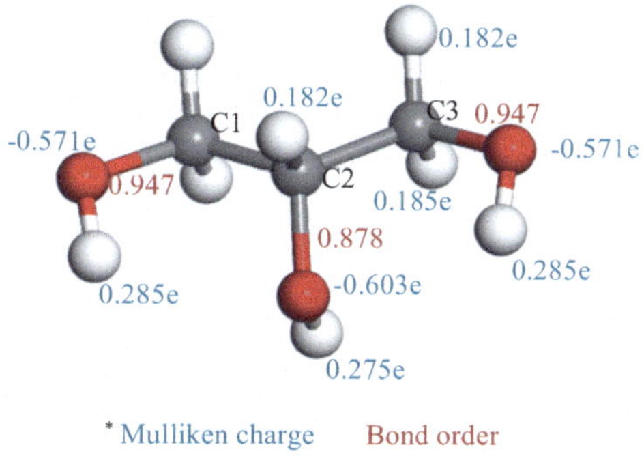

Fig. 9. Mulliken charge population and bond order of the glycerol molecule

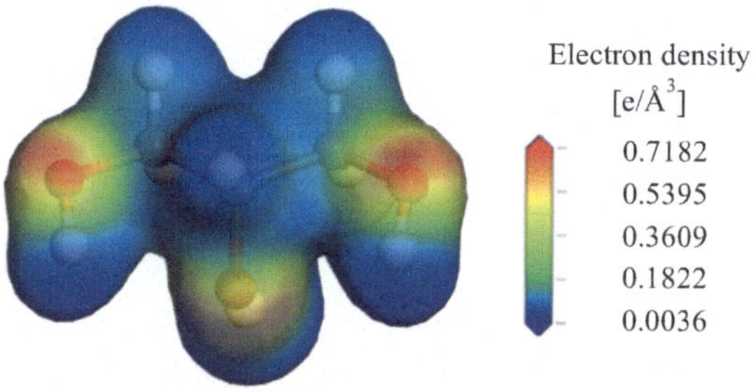

Fig. 10. Spatial distribution of the electron densities of the glycerol molecule

by comparing Reactions (10) and (11), which is, the rate-determining step in the hydrogenation and hydrolysis of the primary hydroxyl group of glycerol to produce 1,2-PDO is a protonation reaction. Similarly, through comparing Reactions (12) and (13), Reactions (14) and (15), Reactions (16) and (17), Reactions (18) and (19), the rate-determining steps in the hydrogenation and hydrolysis of glycerol to either 1,3-PDO or 1,2-PDO are protonation reactions, whether it is in the Brønsted acid site at the meta position of Pt and in the Brønsted acid site without Pt loading. In other words, the protonation reaction is an important reaction that affects the glycerol conversion rate and selectivity in glycerol hydrogenation and hydrolysis, and it should be the research focus.

Next, the protonation reactions of the hydroxyl groups in different positions of glycerol are studied in order to find out the effect of Pt loading on product selectivity. In the Brønsted acid site at the ortho position of Pt, the energy barrier to protonate the intermediate secondary hydroxyl group of glycerol is 36.322 kcal/mol (Reaction (8)), and the energy barrier to protonate the primary hydroxyl group of glycerol is 44.092 kcal/mol (Reaction (10)). The energy barrier to protonate the secondary hydroxyl group is 7.76 kcal/mol lower than the energy barrier to protonate the primary hydroxyl group. Name that, the Brønsted acid site in the ortho position of Pt is easier to activate the secondary hydroxyl group of glycerol to generate 1,3-PDO, and 1,3-PDO has been increased selectivity. In the Brønsted acid site at the meta position of Pt, the energy barrier to protonate the intermediate secondary hydroxyl group of glycerol is 44.666 kcal/mol

Table 1. The activation energy and enthalpy for the reactions of glycerol at different acid sites

Acid sites	Reaction	Ea/ [kcal/mol]	ΔH/ [kcal/mol]
	(8) 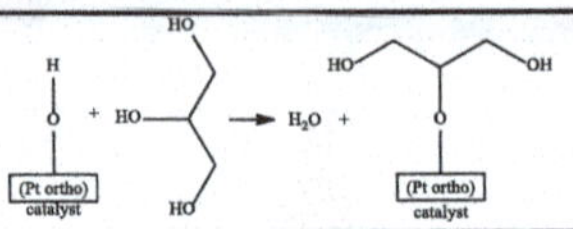	36.332	23.859
Bronsted acid site Pt ortho position 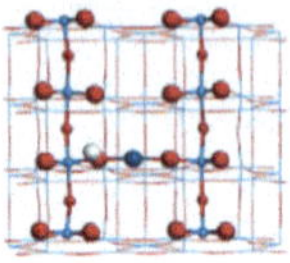	(9) 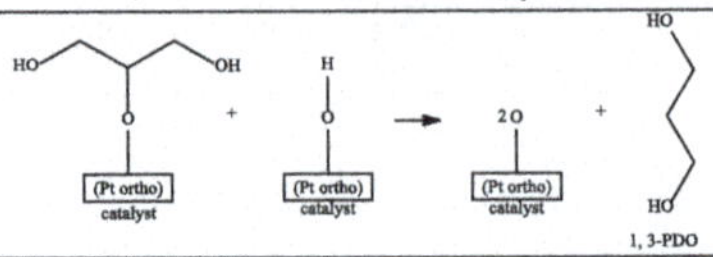	22.407	-4.867
	(10) 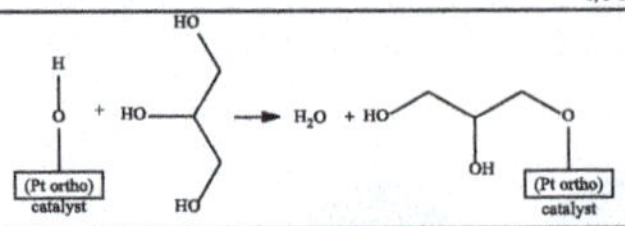	44.092	23.514
	(11) 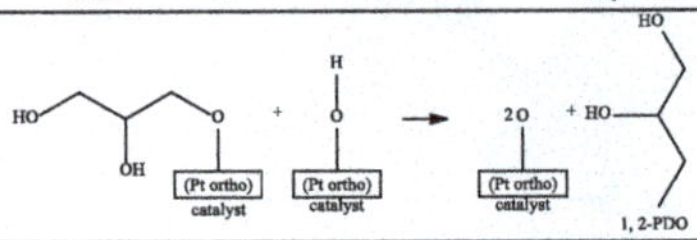	21.207	-5.221
	(12) 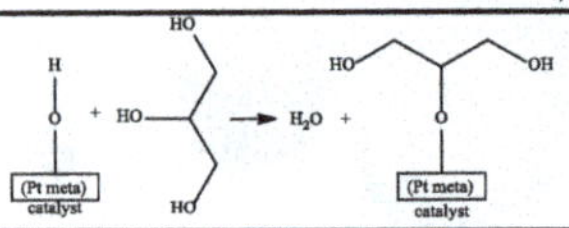	44.666	39.271
Bronsted acid site Pt meta position 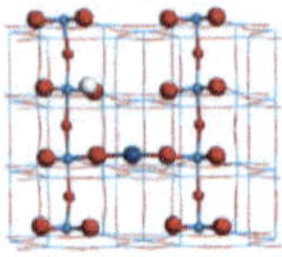	(13) 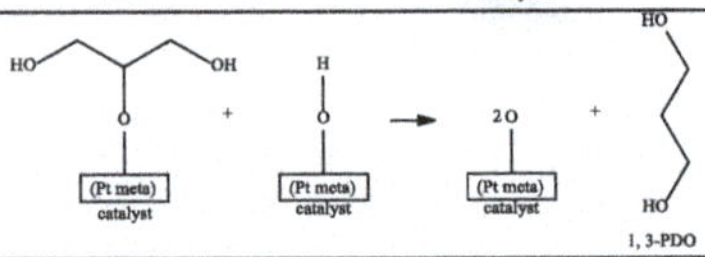	28.505	9.186
	(14) 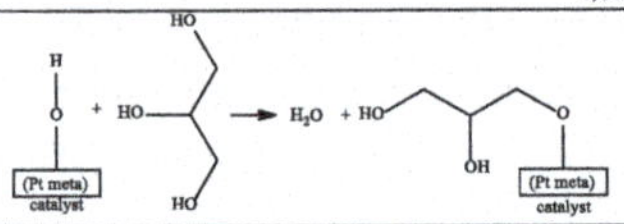	45.037	39.282
	(15) 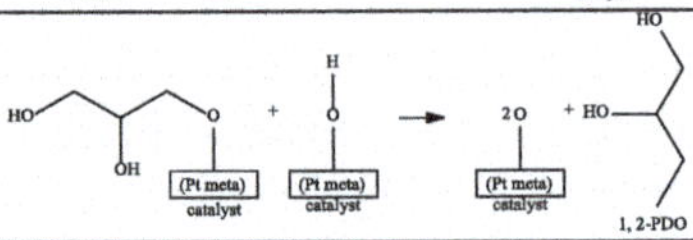	21.648	8.825
	(16) 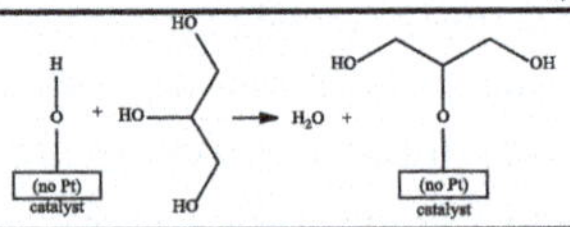	47.166	41.7
Bronsted acid site no Pt 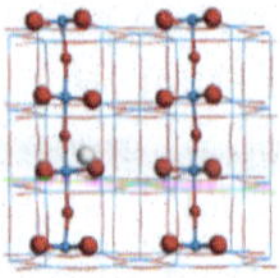	(17) 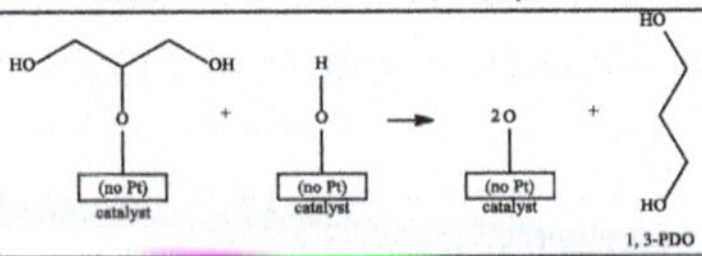	30.685	14.85
	(18) 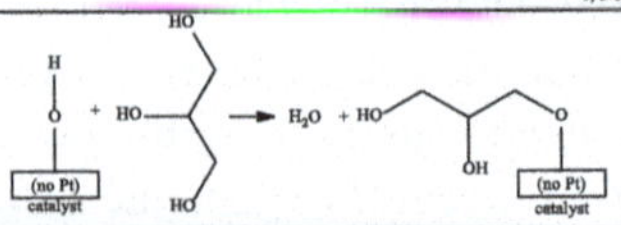	45.914	41.558
	(19) 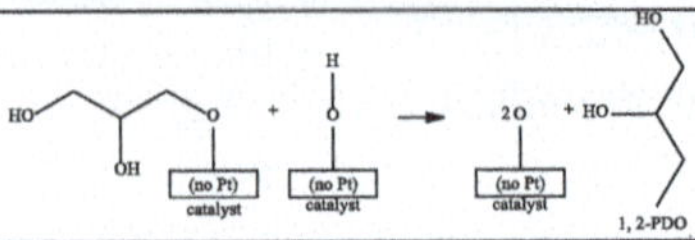	25.624	12.485

(Reaction (12)), and the energy barrier to protonate the primary hydroxyl group of glycerol is 45.037 kcal/mol (Reaction (14)). The energy barrier to protonate the secondary hydroxyl group is only 0.371 kcal/mol lower than the energy barrier to protonate the primary hydroxyl group, significantly less than the 7.76 kcal/mol in the Brønsted acid site at the ortho position of Pt. Therefore, although the Brønsted acid site at the meta position of Pt is easy to activate the secondary hydroxyl group of glycerol to produce 1,3-PDO, 1,3-PDO has lower selectivity as compared to the Brønsted acid site at the ortho position of Pt. In contrast, in the Brønsted acid site not modified by Pt, the energy barrier to protonate the secondary hydroxyl group of glycerol is 47.166 kcal/mol (Reaction (16)), and the energy barrier to protonate the primary hydroxyl group of glycerol is 45.914 kcal/mol (Reaction (18)). Different from the Pt-modified Brønsted acid site, in the Brønsted acid site not modified by Pt, the energy barrier to protonate the primary hydroxyl group is 1.252 kcal/mol, which is lower than the energy barrier to protonate the secondary hydroxyl group. In other words, the Brønsted acid site without Pt modifications is easier to activate the primary hydroxyl group of glycerol, thus producing 1,2-PDO. Hence, Pt modifications are essential to catalyze the hydrogenation and hydrolysis of glycerol to produce 1,3-PDO. Overall, Pt modifications make the product 1,3-PDO more selective than 1,2-PDO, and the Brønsted acid site at the ortho position of Pt has the best selectivity to the product 1,3-PDO.

Last but not least, the roles of different Brønsted acid sites in activation of secondary hydroxyl groups of glycerol are studied, and the effects of Pt on the conversion yield of 1,3-PDO produced by glycerol hydrogenation and hydrolysis are obtained. Through comparing Reaction (8), (12) and (16), clearly the energy barrier to activate glycerol is the lowest in the Brønsted acid site at the ortho position of Pt, followed by that in the Brønsted acid site at the meta position of Pt, and it is the highest in the Brønsted acid site without Pt modifications, denoted by E_a(Reaction (8))<E_a(Reaction (12))<E_a(Reaction (16)). That is to say, glycerol is the easiest to activate and should have the largest conversion rate in the Brønsted acid site at the ortho position of Pt, while the conversion rate will be decreased accordingly in the Brønsted acid site at the meta position of Pt and without Pt modifications. To increase the conversion rate of glycerol to 1,3-PDO by hydrogenation and hydrolysis, brønsted acid sites at the ortho position of Pt should be produced as much as possible during the preparation of catalysts.

In accordance with the above calculations, the first step for catalyzing glycerol hydrogenation on Pt/WO_3 is the protonation of glycerol molecules, in which the catalyst Brønsted acid site provides hydrogen protons (an empty orbital) and the glycerol molecule provides electrons. Pt reduces the activation energy of the adjacent Brønsted acid to catalyze the protonation of glycerol and enhances the selectivity to the hydrogenation of the intermediate secondary hydroxyl group. To find out the reason, the frontier orbitals of the glycerol molecule and the Brønsted acid site are calculated, as shown in Fig. 11. The highest occupied molecular orbital (HOMO) of the glycerol molecule is as shown in Fig. 11a, and its energy order is −7.397eV. The LUMO of the Brønsted acid site at the ortho position of Pt is as shown in Fig. 11a, and its energy order is −7.397eV, which has the smallest energy level difference with glycerol HOMO and is most likely to induce electron transfer. The LUMO in the Brønsted acid site at the meta position of Pt and the LUMO in the Brønsted acid site without Pt modifications are shown in Fig. 11c, d, respectively. Their energy level difference with the glycerol HOMO is larger than that between LUMO in the Brønsted acid site at the ortho position of Pt and the glycerol molecule. Therefore, compared with the Brønsted acid site at the ortho position of Pt, it would be hard for the Brønsted acid site at the meta position of Pt and the Brønsted acid site without Pt modifications to transfer electrons to the glycerol molecule, so the catalytic effect is relatively weak. In the meantime, it can be found from the shape of the orbital of the glycerol HOMO and LUMO in the Brønsted acid site at the meta position of Pt that they have the perfect symmetry, which is the orbitals are close to each other as per the superposition principles of positive and positive, negative and negative. Hence, the introduction of Pt can strengthen the directional activation of the intermediate secondary hydroxyl group by the catalyst.

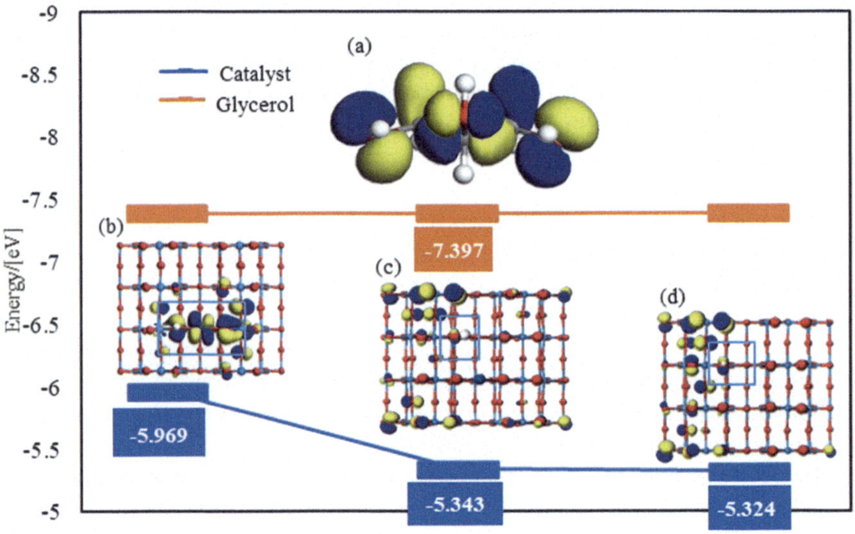

Fig. 11. Frontier orbitals of glycerol and the Brønsted acid sites. Yellow (+) is α spin; blue (–) is β spin. (a) The HOMO of the glycerol molecule, (b) The LUMO of the Brønsted acid site at the ortho position of Pt, (c) The LUMO of the Brønsted acid site at the meta position of Pt, (d) The LUMO of the Brønsted acid site without Pt modifications

4 Conclusions

The Pt/WO_3 catalyzed hydrogenation of glycerol to prepare 1,3-PDO system was studied through DFT theoretical calculation. The C—OH bond in the intermediate glycerol is energetically easier to break than the C—OH bond on both ends; the O atom in the OH is electron-rich, and the negative charge of the secondary hydroxyl O atom in the middle is greater than that of the primary hydroxyl O atoms at both ends, and it is easier to interact with the Brønsted acid site, but steric hindrance is greater at the secondary hydroxyl group in the middle. Pt on the WO_3 surface enhances the acid strength of the Brønsted acid site within 6Å, and there is the most obvious increase in acid strength especially in the Brønsted acid site at the ortho position. In the Brønsted acid site far away from Pt, the acid strength decreases slightly and H coverage is increased accordingly, which is conducive to forming the Brønsted acid site in the vicinity of Pt. Pt reduces the activation energy of the active phase, reduces the activation energy for glycerol hydrogenation, and enhances the selectivity to the hydrogenation of the intermediate secondary hydroxyl group. There is a high match of orbitals between the Brønsted acid sites in the vicinity of Pt on the WO_3 surface and the intermediate secondary hydroxyl group of the glycerol molecule. The small energy level difference between frontier orbitals reduces the activation energy to protonate the intermediate secondary hydroxyl group of glycerol and increases the 1,3-PDO selectivity. To gain higher selectivity to 1,3-PDO, the appropriate combination of Pt loading and Brønsted acid sites should be considered in the design of catalysts, so as to selectively activate the intermediate secondary hydroxyl group of glycerol and increase the yields and selectivity to 1,3-PDO. Through theoretical chemistry calculations, it provides a theoretical basis and indicates the direction of catalyst design for the development of catalysts with high yield and low cost, so that the main product of glycerol hydrogenation reaction is 1,3-PDO and relatively high yield can be obtained. Through this study, a more environmentally friendly and economical route has been provided to alleviate the problem of glycerol overcapacity and improve the production of 1,3-PDO intermediates using traditional industrial methods, which are highly toxic and harmful to human health. This provides basic data for the industrial application of Pt/WO_3 catalytic hydrogenation of glycerol to prepare 1,3-propanediol.

Acknowledgments. This work was supported by Sinopec Research Institute of Petroleum Processing Project (PR20230117) and Inter-governmental international science and technology innovation cooperation projects (2021YFE0191200).

References

1. Xue, L.: Synthesis of Hierarchically Porous SAPO-11 Zeolite for the Hydrogenolysis of Glycerol to 1,2-Propanediol. Southeast University (2019)
2. Salvi, B.L., Panwar, N.L.: Biodiesel resources and production technologies–a review. Renew. Sustain. Energy Rev. **16**, 3680–3689 (2012)
3. Vivek, N., Pandey, A., Binod, P.: 31—Production and applications of 1,3-Propanediol. In: Pandey, A., Negi, S., Soccol, C.R. (eds.) Current Developments in Biotechnology and Bioengineering, pp. 719–738. Elsevier (2017)
4. Salazar, J.B., Falcone, D.D., Pham, H.N., Datye, A.K., Passos, F.B., Davis, R.J.: Selective production of 1,2-propanediol by hydrogenolysis of glycerol over bimetallic Ru–Cu nanoparticles supported on TiO_2. Appl. Catal. A **482**, 137–144 (2014)
5. Yang, M., et al.: Genetically engineered strains: Application and advances for 1,3-Propanediol production from glycerol. Food Technol. Biotech. **56**, 3–15 (2018)
6. Sun, Y.Q., et al.: Advances in bioconversion of glycerol to 1,3-propanediol: prospects and challenges. Process Biochem. **71**, 134–146 (2018)
7. Ma, Y., et al.: Designing effective homogeneous catalysis for glycerol valorisation: selective synthesis of a value-added aldehyde from 1,3-propanediol via hydrogen transfer catalysed by a highly recyclable, fluorinated Cp*Ir(NHC) catalyst. Catal. Today **307**, 248–259 (2018)

8. Hilaly A.K.S.I.: Method of recovering 1,3-propanediol from fermentation broth (2002)
9. Nakamura, G., Hsu.: Method for the production of 1,3-propanediol by recombinant microorganisms (1997)
10. Brar, S.K., Sarma, S.J., Pakshirajan, K.: Platform Chemical Biorefinery: Future Green Chemistry. Elsevier, Amsterdam (2017)
11. Pandey, A., Negi, S., Soccol, C.R.: Current Developments in Biotechnology and Bioengineering: Production, Isolation and Purification of Industrial Products. Elsevier (2017)
12. Varghese, J.J., et al.: Synergistic contribution of the acidic metal oxide-metal couple and solvent environment in the selective hydrogenolysis of glycerol: a combined experimental and computational study using ReO_x–Ir as the catalyst. ACS Catal. **9**, 485–503 (2019)
13. Da Silva Ruy, A.D., de Brito Alves, R.M., Reis Hewer, T.L., de Aguiar Pontes, D., Gomes Teixeira, L.S., Magalhães Pontes, L.A.: Catalysts for glycerol hydrogenolysis to 1,3-propanediol: A review of chemical routes and market. Catal. Today (2020)
14. Samudrala, S.P.: Glycerol Transformation to Value-Added 1,3-Propanediol Production: A Paradigm for a Sustainable Biorefinery Process (2019)
15. Feng, J., Xu, B., Liu, D.R., Xiong, W., Wang, J.B.: Production of 1,3-Propanediol by catalytic hydrogenolysis of glycerol. Adv. Mater. Res. 791–793, 16–19 (2013)
16. Tomishige, K., Nakagawa, Y., Tamura, M.: Production of diols from biomass. In: Fang, Z., Smith, J.R.L., Qi, X. (eds.) Production of Platform Chemicals from Sustainable Resources, pp. 343–373. Springer, Singapore (2017)
17. Chaminand, J., Djakovitch, L., Gallezot, P., Marion, P., Pinel, C., Rosier, C.C.: Glycerol hydrogenolysis on heterogeneous catalysts. Green Chem. **6**, 359–361 (2004)
18. Kusunoki, Y., Miyazawa, T., Kunimori, K., Tomishige, K.: Highly active metal–acid bifunctional catalyst system for hydrogenolysis of glycerol under mild reaction conditions. Catal. Commun. **6**, 645–649 (2005)
19. Miyazawa, T., Kusunoki, Y., Kunimori, K., Tomishige, K.: Glycerol conversion in the aqueous solution under hydrogen over Ru/C + an ion-exchange resin and its reaction mechanism. J. Catal. **240**, 213–221 (2006)
20. Sun, D., Yamada, Y., Sato, S., Ueda, W.: Glycerol hydrogenolysis into useful C3 chemicals. Appl. Catal. B **193**, 75–92 (2016)
21. Zhou, W., et al.: WO_x domain size, acid properties and mechanistic aspects of glycerol hydrogenolysis over $Pt/WO_x/ZrO_2$. Appl. Catal. B **242**, 410–421 (2019)
22. Liu, L., Kawakami, S., Nakagawa, Y., Tamura, M., Tomishige, K.: Highly active iridium–rhenium catalyst condensed on silica support for hydrogenolysis of glycerol to 1,3-propanediol. Appl. Catal. B **256**, 117775 (2019)
23. Nakagawa, Y., Tamura, M., Tomishige, K.: Catalytic materials for the hydrogenolysis of glycerol to 1,3-propanediol. J. Mater. Chem. A **2**, 6688–6702 (2014)
24. Wan, X., et al.: Interface synergy between IrO_x and H-ZSM-5 in selective C–O hydrogenolysis of glycerol toward 1,3-propanediol. J. Catal. **375**, 339–350 (2019)
25. Johnson, D.T., Taconi, K.A.: The glycerin glut: Options for the value-added conversion of crude glycerol resulting from biodiesel production. **26**, 338–348 (2007)
26. Wang, K., Hawley, M., DeAthos, S.: Conversion of glycerol to 1,3-Propanediol via selective dehydroxylation. Ind. Eng. Chem. Res. **42**, 2913–2923 (2003)
27. Thomas, H., Armin, N., Dietrich, A., Herbert, K., Walter, G.: Process for the simultaneous production of 1,2- and 1,3-propanediol (1993)
28. Qin, L., Song, M., Chen, C.: Aqueous-phase deoxygenation of glycerol to 1,3-propanediol over $Pt/WO_3/ZrO_2$ catalysts in a fixed-bed reactor. Green Chem. **12**, 1466–1472 (2010)
29. Li, D., Zhou, Z., Qin, J., Li, Y., Liu, Z., Wu, W.: Cu-WO_x-TiO_2 catalysts by modified evaporation-induced self-assembly method for glycerol hydrogenolysis to 1, 3-Propanediol. Chem. Select **3**, 2479–2486 (2018)

30. Kant, A., et al.: Hydrogenolysis of glycerol over Ni, Cu, Zn, and Zr supported on H-beta. Chem. Eng. J. **317**, 1–8 (2017)
31. Huang, L., Zhu, Y., Zheng, H., Ding, G., Li, Y.: Direct conversion of glycerol into 1,3-propanediol over Cu-$H_4SiW_{12}O_{40}/SiO_2$ in vapor phase. Catal. Lett. **131**, 312–320 (2009)
32. Gebretsadik, F.B., Llorca, J., Salagre, P., Cesteros, Y.: Hydrogenolysis of glycidol as an alternative route to obtain 1,3-Propanediol selectively using MO_x-modified nickel-copper catalysts supported on acid mesoporous saponite. ChemCatChem **9**, 3670–3680 (2017)
33. Tamura, M., Amada, Y., Liu, S., Yuan, Z., Nakagawa, Y., Tomishige, K.: Promoting effect of Ru on Ir-ReO_x/SiO_2 catalyst in hydrogenolysis of glycerol. J. Mol. Catal. A: Chem. **388–389**, 177–187 (2014)
34. Amada, Y., Shinmi, Y., Koso, S., Kubota, T., Nakagawa, Y., Tomishige, K.: Reaction mechanism of the glycerol hydrogenolysis to 1,3-propanediol over Ir-ReO_x/SiO_2 catalyst. Appl. Catal. B **105**, 117–127 (2011)
35. Luo, W., Lyu, Y., Gong, L., Du, H., Wang, T., Ding, Y.: Selective hydrogenolysis of glycerol to 1,3-propanediol over egg-shell type Ir-ReO_x catalysts. RSC Adv. **6**, 13600–13608 (2016)
36. Deng, C., Leng, L., Duan, X., Zhou, J., Zhou, X., Yuan, W.: Support effect on the bimetallic structure of Ir-Re catalysts and their performances in glycerol hydrogenolysis. J. Mol. Catal. A: Chem. **410**, 81–88 (2015)
37. Deng, C., Duan, X., Zhou, J., Zhou, X., Yuan, W., Scott, S.L.: Ir-Re alloy as a highly active catalyst for the hydrogenolysis of glycerol to 1,3-propanediol. Catal. Sci. Technol. **5**, 1540–1547 (2015)
38. Nakagawa, Y., Ning, X., Amada, Y., Tomishige, K.: Solid acid co-catalyst for the hydrogenolysis of glycerol to 1,3-propanediol over Ir-ReO_x/SiO_2. Appl. Catal. A **433–434**, 128–134 (2012)
39. Wang, J., Lei, N., Yang, C., Su, Y., Zhao, X., Wang, A.: Effect of promoters on the selective hydrogenolysis of glycerol over Pt/W-containing catalysts. Cuihua Xuebao/Chinese J. Catal. **37**, 1513–1519 (2016)
40. Nakagawa, Y., Shinmi, Y., Koso, S., Tomishige, K.: Direct hydrogenolysis of glycerol into 1,3-propanediol over rhenium-modified iridium catalyst. J. Catal. **272**, 191–194 (2010)
41. Shi, G., et al.: Selective hydrogenolysis of glycerol to 1,3-propanediol over Pt-WO_x/SAPO-34 catalysts. Mol. Catal. **456**, 22–30 (2018)
42. Uttraporn, P., Praserthdam, P.: Effect of calcination temperature and support type of Pt/WO_x/boehmite catalyst on 1,3-propanediol production from hydrogenolysis of glycerol. IOP Conf. Ser. Mater. Sci. Eng. **559**, 12012 (2019)
43. Liu, L., Zhang, Y., Wang, A., Zhang, T.: Mesoporous WO_3 supported Pt catalyst for hydrogenolysis of glycerol to 1,3-propanediol. Chinese J. Catal. **33**, 1257–1261 (2012)
44. Shi, G., Cao, Z., Xu, J., Jin, K., Bao, Y., Xu, S.: Effect of WO_x doping into Pt/SiO_2 catalysts for glycerol hydrogenolysis to 1,3-Propanediol in liquid phase. Catal. Lett. **148**, 2304–2314 (2018)
45. Zhu, S., et al.: Alkaline metals modified Pt-$H_4SiW_{12}O_{40}/ZrO_2$ catalysts for the selective hydrogenolysis of glycerol to 1,3-propanediol. Appl. Catal. B **140–141**, 60–67 (2013)
46. Yang, M., et al.: Pt/Nb-WO_x for the chemoselective hydrogenolysis of glycerol to 1,3-propanediol: Nb dopant pacifying the over-reduction of WO_x supports. Cuihua Xuebao/Chinese J. Catal. **39**, 1027–1037 (2018)
47. Kurosaka, T., Maruyama, H., Naribayashi, I., Sasaki, Y.: Production of 1,3-propanediol by hydrogenolysis of glycerol catalyzed by Pt/WO_3/ZrO_2. Catal. Commun. **9**, 1360–1363 (2008)
48. Gong, L., et al.: Solvent effect on selective dehydroxylation of glycerol to 1,3-propanediol over a Pt/WO_3/ZrO_2 catalyst. Cuihua Xuebao/Chinese J. Catal. **30**, 1189–1191 (2009)
49. Ten Dam, J., Djanashvili, K., Kapteijn, F., Hanefeld, U.: Pt/Al_2O_3 Catalyzed 1,3-Propanediol formation from glycerol using tungsten additives. ChemCatChem **5**, 497–505 (2013)

50. Edake, M., Dalil, M., Darabi Mahboub, M.J., Dubois, J.L., Patience, G.S.: Catalytic glycerol hydrogenolysis to 1,3-propanediol in a gas-solid fluidized bed. RSC Adv. **7**, 3853–3860 (2017)
51. Yang, C., et al.: Understanding the promotional effect of Au on Pt/WO_3 in hydrogenolysis of glycerol to 1,3-propanediol. Cuihua Xuebao/Chinese J. Catal. **39**, 1366–1372 (2018)
52. Wang, J., et al.: Hydrogenolysis of Glycerol to 1,3-propanediol under low hydrogen pressure over WO_x-supported single/pseudo-single atom Pt catalyst. Chemsuschem **9**, 784–790 (2016)
53. Wu, Z., et al.: One-pot catalytic production of 1,3-propanediol and γ-valerolactone from glycerol and levulinic acid. Catal. Today **302**, 217–226 (2018)
54. Gu, M., et al.: The effect of catalytic structure modification on hydrogenolysis of glycerol into 1,3-Propanediol over platinum nanoparticles and ordered mesoporous alumina assembled catalysts. Ind. Eng. Chem. Res. **56**, 13572–13581 (2017)
55. Zhu, S., Qiu, Y., Zhu, Y., Hao, S., Zheng, H., Li, Y.: Hydrogenolysis of glycerol to 1,3-propanediol over bifunctional catalysts containing Pt and heteropolyacids. Catal. Today **212**, 120–126 (2013)
56. García-Fernández, S., et al.: New approaches to the $Pt/WO_x/Al_2O_3$ catalytic system behavior for the selective glycerol hydrogenolysis to 1,3-propanediol. J. Catal. **323**, 65–75 (2015)
57. Lei, N., et al.: Effective hydrogenolysis of glycerol to 1,3-Propanediol over metal-acid concerted $Pt/WO_x/Al_2O_3$ catalysts. ChemCatChem **11**, 3903–3912 (2019)
58. García-Fernández, S., Gandarias, I., Requies, J., Soulimani, F., Arias, P.L., Weckhuysen, B.M.: The role of tungsten oxide in the selective hydrogenolysis of glycerol to 1,3-propanediol over $Pt/WO_x/Al_2O_3$. Appl. Catal. B **204**, 260–272 (2017)
59. Zhu, S., Zhu, Y., Hao, S., Chen, L., Zhang, B., Li, Y.: Aqueous-phase hydrogenolysis of glycerol to 1,3-propanediol over $Pt\text{-}H_4SiW_{12}O_{40}/SiO_2$. Catal. Lett. **142**, 267–274 (2012)
60. Feng, S., Zhao, B., Liang, Y., Liu, L., Dong, J.: Improving selectivity to 1,3-Propanediol for glycerol hydrogenolysis using W- and Al-incorporated SBA-15 as support for Pt nanoparticles. Ind. Eng. Chem. Res. **58**, 2661–2671 (2019)
61. García-Fernández, S., Gandarias, I., Tejido-Núñez, Y., Requies, J., Arias, P.L.: Influence of the support of bimetallic platinum tungstate catalysts on 1,3-propanediol formation from glycerol. ChemCatChem **9**, 4508–4519 (2017)
62. Aihara, T., Kobayashi, H., Feng, S., Miura, H., Shishido, T.: Effect of WO_3 loading on the activity of $Pt/WO_3/Al_2O_3$ Catalysts in selective hydrogenolysis of glycerol to 1, 3-Propanediol. Chem. Lett. **46**, 1497–1500 (2017)
63. Priya, S.S., Kumar, V.P., Kantam, M.L., Bhargava, S.K., Chary, K.V.R.: Vapour-phase hydrogenolysis of glycerol to 1,3-propanediol over supported pt catalysts: the effect of supports on the catalytic functionalities. Catal. Lett. **144**, 2129–2143 (2014)
64. Priya, S.S., Kumar, V.P., Kantam, M.L., Bhargava, S.K., Chary, K.V.R.: Catalytic performance of $Pt/AlPO_4$ catalysts for selective hydrogenolysis of glycerol to 1,3-propanediol in the vapour phase. RSC ADV. **4**, 51893–51903 (2014)
65. Priya, S.S., Kumar, V.P., Kantam, M.L., Bhargava, S.K., Srikanth, A., Chary, K.V.R.: High efficiency conversion of glycerol to 1,3-Propanediol using a novel platinum-tungsten catalyst supported on SBA-15. Ind. Eng. Chem. Res. **54**, 9104–9115 (2015)
66. Mizugaki, T., Yamakawa, T., Arundhathi, R., Mitsudome, T., Jitsukawa, K., Kaneda, K.: Selective hydrogenolysis of glycerol to 1,3-propanediol catalyzed by Pt nanoparticles-AlO_x/WO_3. Chem. Lett. **41**, 1720–1722 (2012)
67. Zhu, S., Gao, X., Zhu, Y., Li, Y.: Promoting effect of WO_x on selective hydrogenolysis of glycerol to 1,3-propanediol over bifunctional $Pt\text{-}WO_x/Al_2O_3$ catalysts. J. Mol. Catal. A: Chem. **398**, 391–398 (2015)
68. Zhu, M., Chen, C.: Hydrogenolysis of glycerol to 1,3-propanediol over $Li_2B_4O_7$-modified tungsten–zirconium composite oxides supported platinum catalyst. React. Kinet. Mech. Catal. **124**, 683–699 (2018)

69. Priya, S.S., Bhanuchander, P., Kumar, V.P., Dumbre, D.K., Periasamy, S.R., Bhargava, S.K., Lakshmi Kantam, M., Chary, K.V.R.: Platinum supported on H-Mordenite: a highly efficient catalyst for selective hydrogenolysis of glycerol to 1,3-Propanediol. ACS Sustain. Chem. Eng. **4**, 1212–1222 (2016)
70. Fan, Y., et al.: Pt–WO_x on monoclinic or tetrahedral ZrO_2: crystal phase effect of zirconia on glycerol hydrogenolysis to 1,3-propanediol. Appl. Catal. B **217**, 331–341 (2017)
71. Priya, S.S., Bhanuchander, P., Kumar, V.P., Bhargava, S.K., Chary, K.V.R.: Activity and selectivity of platinum-copper bimetallic catalysts supported on mordenite for glycerol hydrogenolysis to 1,3-propanediol. Ind. Eng. Chem. Res. **55**, 4461–4472 (2016)
72. Oh, J., Dash, S., Lee, H.: Selective conversion of glycerol to 1,3-propanediol using Pt-sulfated zirconia. Green Chem. **13**, 2004–2007 (2011)
73. Samudrala, S.P., Kandasamy, S., Bhattacharya, S.: Turning biodiesel waste glycerol into 1,3-propanediol: catalytic performance of sulphuric acid-activated montmorillonite supported platinum catalysts in glycerol hydrogenolysis. SCI REP-UK **8**, 7484 (2018)
74. Arundhathi, R., Mizugaki, T., Mitsudome, T., Jitsukawa, K., Kaneda, K.: Highly selective hydrogenolysis of glycerol to 1,3-propanediol over a boehmite-supported platinum/tungsten catalyst. Chemsuschem **6**, 1345–1347 (2013)
75. Gong, L., et al.: Selective hydrogenolysis of glycerol to 1,3-propanediol over a Pt/WO_3/TiO_2/SiO_2 catalyst in aqueous media. Appl. Catal. ACatal. A **390**, 119–126 (2010)

An Agent-Based Tobacco Town Model to Investigate Impacts of Policy Interventions

Willaim Strachan[1], Ricardo Colasanti[2], Kashif Zia[2(✉)], Yahya Gamal[3], and Alison Heppenstall[2,3]

[1] College of Medical, Veterinary and Life Sciences, University of Glasgow, Glasgow, UK
[2] MRC/CSO Social and Public Health Sciences Unit, University of Glasgow, Glasgow, UK
kashif.zia@glosgow.ac.uk
[3] Urban Big Data Centre, University of Glasgow, Glasgow, UK

Abstract. This paper proposes a data-driven exploratory agent-based model representing a tobacco town. To investigate the effects of policy interventions on accessibility (cost and availability) and consequently, tobacco consumption, the model compares two economically diverse localities of the city of Glasgow namely Glasgow Kelvin (urban rich) and Glasgow Shettleston (urban poor). Several policies were simulated to reduce the density of retailers in the environment: These were (1) retailer density cap—a random reduction at a percentage of the original density (2) retailer and school proximity buffers at varying distances, and (3) a combination. The results showed that as retailer density decreased, overall costs increased with a threshold effect. For both localities of Glasgow, retailer buffers were predicted to be the most effective single policy to increase costs hence reduce accessibility. However, single policies were generally predicted to exhibit relatively mild effects on cost; a combination of policies contextualized to the locality was necessary to increase the cost of purchasing cigarettes to a degree that might alter smoking behavior. This study demonstrates the usefulness of ABMs in predicting the effects of public health policies and highlights the importance of considering the contextual effect of policies on smoking behaviors.

Keywords: Tobacco Town · Agent-based model · Policy interventions · Environmental sustainability

1 Introduction

Tobacco use continues to be one of the largest contributors to the burden of disease and mortality in the UK [1, 2]. It is a significant global threat to public health with the WHO reporting more than eight million tobacco related deaths worldwide every year [3, 4]. The approach of altering the retail availability of both unhealthy and healthy products as a means of creating healthier communities is frequently under investigation [5]. Lowering the prevalence of tobacco use, particularly with effective and overarching policy, continues to be one of the primary goals of governments and policy makers worldwide.

G. Huang et al. (Eds.): ICESD 2024, ESE, pp. 174–187, 2025.
https://doi.org/10.1007/978-3-031-88683-6_13

In addition to personal health, the impact of environmental tobacco smoke (particularly in closed locations)—that is passive smoking—on smokers as well as nonsmokers has significant health implications [6]. Although tobacco smoke deteriorates the open air as well, the impact of indoor air pollution on health [7] is reported widely and tobacco smoke has been recognized [8] as one of the main sources responsible for that. Therefore, in addition to a hazard to individual health, smoking is also an environmental concern related to public health.

In Scotland, smoking is responsible for one in five adult deaths. It is additionally the largest preventable cause of premature death and poor health [2]. Smoking rates continue to decline, however, the rate disparity at different levels of deprivation does not; smoking is consequently becoming an increasing health inequality challenge [9]. Therefore, it is necessary to approach the problem at a locality level instead of a global policy decision.

Health risks are known to increase as smokers continue to use tobacco, but smoking cessation can substantially reduce these risks [10]. With an aim towards a smoke free generation, one suggestion for the future direction of Scottish policy is a focus on the availability of tobacco products [11]. Tobacco use is already heavily regulated in Scotland, with the introduction of the Public Place Smoking Ban in 2006, consolidation of regulation of display, packaging and labelling, alongside rigorous taxation [12]. Retailer density reduction represents a contemporary approach of tobacco harm reduction policy in Scotland and the rest of the world [13, 14]. Current research is focusing on the specific policies that can effectively reduce retailer density, and subsequently whether these reductions will lead to notable public health benefits.

The evidence base supporting the link between the availability of tobacco and tobacco use is substantial [15–17]. Investigating policy options aimed at reducing the density of tobacco retailers has continually been in the scope of public health tobacco research. A cross-sectional study in 2016 investigated the relationship between smoking behavior and tobacco retailer density at home and around schools, in a Scottish context [16]. The study highlighted that living in an area with a higher density of tobacco retailers resulted in a greater likelihood of smoking in adolescents. A 2020 University of Glasgow study aimed to spatially simulate retailer reduction policies in Glasgow. This study highlighted that these policies could significantly reduce the density of tobacco retailers, particularly in areas with higher socio-economic disadvantage [18]. However, the authors also noted potential unintended consequences, such as a clustering of tobacco retailers in areas where they remain permitted; highlighting a significant need for further policy research before reduction policies are implemented.

Individual communities often require tailored interventions and policies to address their unique requirements and address underlying factors [19]. While there is a significant amount of research detailing successful population health strategies, research often fails to consider the dynamic socioecological and cultural context of a particular community or existing policies [20]. Limited resources also often render strategies that have shown success elsewhere unfeasible or unsustainable. It is typically ineffective to randomly choose evidence-based policies and interventions without considering their relevance and appropriateness for a particular context.

Computational modelling in public health research represents a powerful tool for visualizing, analyzing, and predicting complex and dynamic systems without the cost and impracticality of conventional research techniques [21]. Over the last two decades, there has been a notable surge in the use of computational modelling to explore complex systems from land-use and agricultural policy [22], to the management of non-communicable diseases [23].

Agent Based Modelling (ABM) is a specific application of computational modelling that allows experimental policy options and interventions to be rapidly tested in a controlled environment. ABM defines the use of a computational approach to simulate the interactions of agents with specific characteristics, including individuals, households, and governments [21]. These agents adapt their behavior based on their experiences and interactions with other agents and the environment. This approach allows for the emergence of population-level phenomena that are not predictable based solely on individual behaviors [24]. Effective agent-based modelling requires a deep understanding of the system being modelled, including the behavior of individual agents, the structure of the social network, and the environmental factors that influence the system. By incorporating these factors into the model, researchers can create a more accurate representation of the system and better predict how it will behave under different conditions. This allows policymakers to make informed decisions based on the results of the simulations.

ABM has been recognized as a crucial instrument in tobacco control policy and regulation. ABM can assist in creating, executing, and enforcing policies aimed at changing people's behaviors concerning tobacco use [25]. ABM, in the context of tobacco retailer density, is most valuable when applied as a policy testing 'laboratory' to experiment with different policies in silico [26]. Our work is an extension of a seminal work named Tobacco Town. The Tobacco Town Agent Based Model was part of a 2017 study that used complex systems modelling to investigate the potential effects of tobacco retailer reduction policies in an abstract environment of four different Californian town types (suburban/urban rich and poor) [27]. The model analyses how the expense of obtaining cigarettes, which comprise both the direct cost of the product and the additional cost of traveling to a tobacco retailer, varied due to retailer reduction policies. When assessing the policy scenarios, the fundamental aspect explored by the model is the influence of retail density dynamics on overall costs. The foundations of the tobacco town model have been more recently reapplied to Minnesota to investigate menthol cigarette sales [28].

This ongoing research project aims to reproduce and simplify the Tobacco Town agent-based model [27] in NetLogo [29], with the vision to make it publicly accessible as a tool to investigate tobacco retailer policy in different environments. The secondary objective of this project is to investigate the potential of different policy interventions in a Scottish context by using demographic, socioeconomic, and retailer empirical data representing two demographically opposite localities of Glasgow. We also provide an outlook (approach and challenges) of how the model can be extended to include real GIS and population data.

2 The Tobacco Town Model

The Tobacco Town agent-based model is designed to explore the changes in purchase behavior and both direct and indirect purchase costs in response to a variety of tobacco retailer density and placement policy options in the two diverse localities of Glasgow. The model presented in [27] was simplified and contextualized in terms of data used and for the spatial features.

2.1 Model Setup

The model environment consists of a grid of roads and intersections representing two areas of the city of Glasgow; Urban Rich area (Kelvin Constituency) and an Urban Poor area (Shettleston Constituency) each initialized using empirical data scaled appropriately to the size of the grid. The spatial features consisted of workplace, school and retailer density. Agents in the model represented adult smokers and were generated based on the population density in each town type. In the interest of simplicity all agents were generated as active smokers, initiation and cessation were not programmed as an agent function.

In Fig. 1 the initial setup of the model is depicted. All intersections are nodes that can be occupied by smokers, workplaces, retailers, or schools. A data-driven scaled population of smokers (persons), workplaces (blue boxes), retailers (orange targets), and schools (white trees) randomly occupying these nodes. The retailers are scaled in size to show the variation in average prices of cigarettes at a retailer. The smokers are scaled in size to show the variation in wages per hour. The smokers are also colored differently based on transport mode; red if car, yellow if cycling (other), and green if walking.

2.2 Data

A summary of data sources used is given in Table 1. The majority of the empirical data for Glasgow city was obtained from the official census data of Scotland. Other sources were also utilized. However, the data for density and price of specific tobacco retailer types was not available for Glasgow—point of sale data for convenience stores [30] was used as an approximation for all retailers based on the reported distribution in [31]. Wages data (not shown in the table) was also taken from official sources for both localities of the urban town types as a close approximation. The simulation setup shown in Fig. 1 is a scaled-down environment created based on this data with random placements.

2.3 Model Behavior

Each simulated day consists of two periods (morning and evening) in which agents conduct different tasks. In the morning period agents make a purchase decision—if their current inventory is less than their assigned smoking rate—and then commute from home to work potentially purchasing tobacco enroute.

In line with [27] and with standard economic theory it is assumed that individuals have complete information about both direct and indirect expenses, including the price

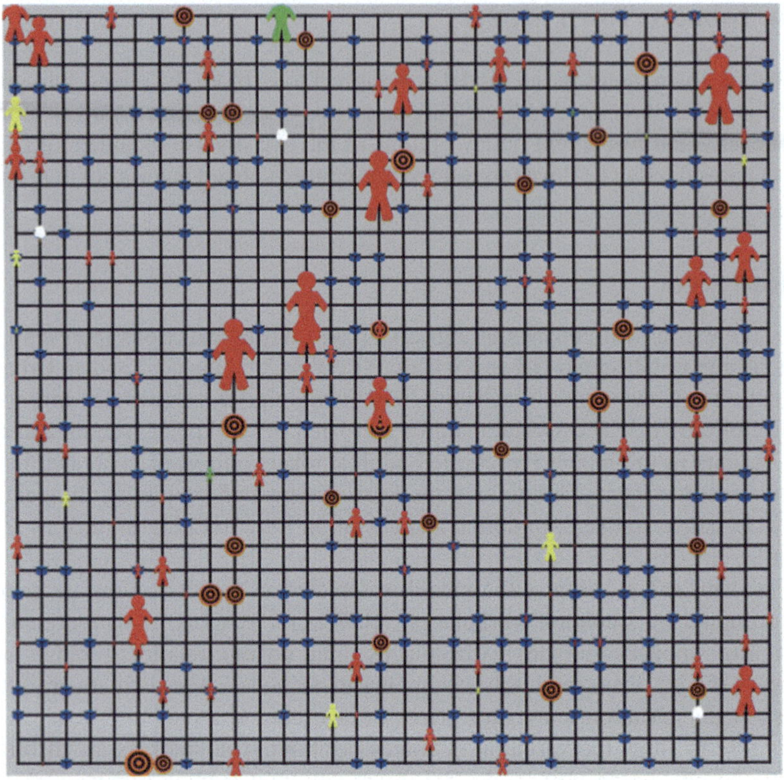

Fig. 1. Model setup showing smokers (persons), workplaces (boxes), retailers (targets), and schools (trees) lying at intersecting nodes.

of cigarettes at every retailer and the expense involved in altering their commute route to buy cigarettes. With this information agents make choices about where to buy cigarettes and the quantity to purchase to minimize the overall cost per pack.

The primary result the model produces is a theoretical total expense combining the time and travel cost to get to a tobacco retailer and the monetary cost of cigarettes. Smokers each had six time-invariant attributes of smoking rate, home and work locations, wage, transport type of car, bike or walk, and a discount rate that devalues packs bought in the future to prevent agents purchasing an infinite number of packs.

Agents choose an optimal retailer to purchase from and an optimal quantity of packs to purchase using the rational decision rules (reported as Appendix A in [27]). For each retailer in the model, the agents first compute the optimal quantity of cigarettes that can be bought (with a particular focus on the discounting factor). This internally depends on the expected cost of purchasing that quantity. This generates the expected cost of purchasing an optimal quantity of cigarettes at each retailer. One of these retailers is then selected using the trembling hand process (see Appendix A in [27] for further details).

Table 1. Table of Glasgow Empirical Data and sources.

Town type Input	Urban poor	Urban rich	Source/Notes
Glasgow Area	*East End (Shettleston)*	*City and West (Kelvin)*	Represent constituencies
			Densities are per square mile
School density	2.82	3.64	School Catchments Open Data 2021
Workplace density	72.81	219.62	LuminoCity3D and Scotland's Business Base
Population density	9504	18037	SIMD Data 2020
Area	5.76	5.46	GIS SIMD Data 2020
Transport mode			2011 Scottish Census
Car proportion	0.477		*"Other" category includes all types of public transport and cycling; the proportions do not sum to 1 because work from home is not modelled*
Walk proportion	0.121		
Other proportion	0.303		
Transport speed (mph)			*Invariant across area types (Bike and Walk taken from Tobacco Town)*
Car	19.9	19.9	Tom Tom 2023
Bike	7.5	7.5	Allen et al. 1998
Walk	2.1	2.1	Knoblauch et al. 1996
Efficiency (mpg)	33		Department of Transport 2019
Fuel Price (£/Gallon)	6.62		National Stats weekly avg. Fuel price (2023)

A potential difference between this model and the original is the use of a node network to represent the roads and intersections of the environment. The network is generated by creating nodes and then creating links between adjacent nodes to represent the network seen in Fig. 1. This permits the model to make use of the NetLogo network extension that allows straightforward calculation of the distance between two points on the grid of nodes, i.e., the shortest distance from home to work, as opposed to programming pathfinding algorithms.

Table 2. Density Reduction and Cost Increase Results | Glasgow Tobacco Town

BASELINE	Density (Per mi^2) 16.53	Cost (£) 8.26		Density (Per mi^2) 28.1	Cost (£) 8.57	
POLICY		[95% CI]	% Change		[95% CI]	% Change
Density Cap						
90%	15.08	0.08 (0.04, 0.21)	0.97%	25.41	0.01 (–0.14, 0.11)	0.12%
80%	13.43	0.12 (0.01, 0.27)	1.45%	22.52	0.01 (–0.01, 0.15)	0.12%
70%	11.78	0.13 (0.04, 0.18)	1.57%	19.83	0.06 (0.02, 0.19)	0.7%
60%	9.92	0.16 (0.04, 0.27)	1.94%	16.94	0.13 (0.01, 0.25)	1.52%
50%	8.26	0.23 (0.14, 0.30)	2.78%	14.05	0.13 (0.09, 0.29)	1.52%
Retailer Buffer						
500Ft	12.27	0.18 (0.01, 0.34)	2.18%	16.28	0.06 (–0.21, 0.36)	0.7%
1000Ft	7.15	0.22 (0.14, 0.29)	2.66%	8.26	0.22 (0.1, 0.36)	2.57%
1500Ft	3.76	0.38 (0.22, 0.54)	4.6%	3.93	0.49 (0.15, 0.83)	5.72%
School Buffer						
500Ft	14.59	0.02 (–0.04, 0.08)	0.24%	24.01	0.06 (–0.08, 0.20)	0.7%
1000Ft	11.61	0.09 (–0.003, 0.02)	1.09%	18.6	0.07 (0.02, 0.23)	0.82%

(continued)

Table 2. *(continued)*

BASELINE POLICY	Density (Per mi^2) 16.53	Cost (£) 8.26 [95% CI]	% Change	Density (Per mi^2) 28.1	Cost (£) 8.57 [95% CI]	% Change
1500Ft	7.36	0.2 (0.07, 0.33)	2.42%	10.99	0.13 (0.07, 0.32)	1.52%
Combination						
Moderate	4.92	0.22 (0.02, 0.42)	2.66%	5.83	0.24 (0.12, 0.35)	2.8%
High	2.12	0.63 (0.38, 0.87)	7.63%	2.77	0.55 (0.39, 0.71)	6.42%

Glasgow Urban Poor (Shettleston) Glasgow Urban Rich (Kelvin)

The original model found the shortest distance between a selected retailer and a point on the predefined route from home to work, the agents then follow this route off the normal path to the outlet and then back to their original path to continue commuting. Our model finds the difference between the normal commute and an optimal commute from home to the retailer and then from the retailer to work. This was deemed a simplified and potentially more realistic pathfinding method that theoretically might yield slightly lower travel costs than the original model but should still produce the same relative increase in cost when policies are implemented.

3 Simulation and Results

A key criterion of this NetLogo model was to allow baseline agent and environmental parameters to be easily altered so that the model can be used to test policies in many different environments. Four different policies were included with the option to test in isolation, and in combination, at varying strengths. (1) Retailer density limits—90, 80, 70, 60, 50% of initial density. (2) School proximity buffers at 500, 1000 and 1500ft (3) Retailer proximity buffers at 500, 1000 and 1500ft, and (4) Two-level qualitative combination.

Each simulated environment and policy combination was run 5 times to ensure stable agent behavior. During these runs, data was collected on all components of the model in the form of aggregate data on total direct and indirect costs, distance, time for travel, purchase price, and quantity purchased. The retailer density at the end of each run was also collected.

As expected in a nutshell, it was observed that as retailer density decreases, overall costs increase. This relationship exhibits a threshold effect at approximately 3 retailers per square mile, whereby at densities lower than three, costs begin to increase significantly. Next, we present the simulation results in more detail.

The results of the Glasgow model runs are included in Table 2, including the change in retailer density and the relative increase in the overall cost of purchasing cigarettes for the different types of policy interventions across the two areas. We categorize the results across the effects of retailers' density and cost, application caps on retailers, application of buffer spaces, and a qualitative combination of these representing strict and moderate policy.

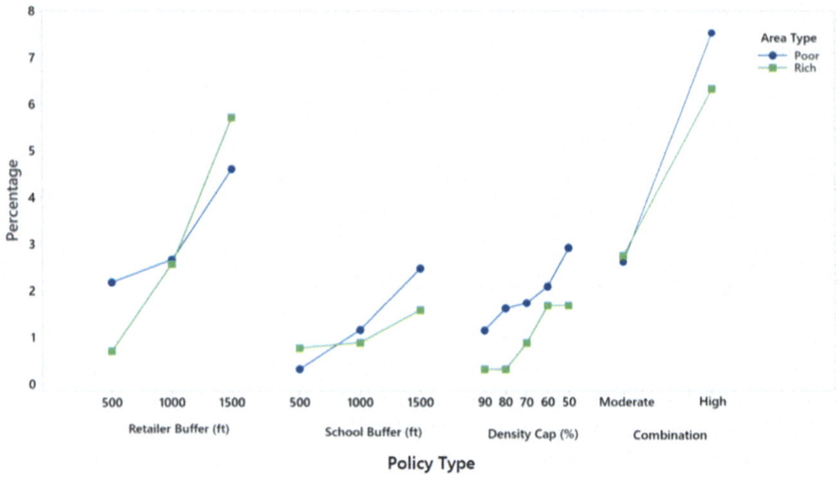

Fig. 2. Relative Overall Travel Plus Purchase Cost Increase by Policy Type

Density reduction vs. Effect on cost: Before policies were simulated, the number of retailers per square mile ranged from 16.54 (for Urban Poor) to 28.1 (for Urban Rich) in the baseline runs. The model predicted that greater reductions in retailer density were achieved with higher intensity interventions. For instance, implementing a 500-foot retailer buffer reduced retailer density from 16.54 to 12.27 in urban poor areas, while a much larger buffer of 1500 feet decreased the density to 3.76 retailers per square mile.

Table 2 presents the average increase in cost per pack of cigarettes—including travel and purchase cost—for each policy strategy. Figure 2 illustrates the impact of each policy strategy on the total cost of cigarettes. Each line graph in the figure depicts the percentage increase in cost relative to the baseline costs. The results showed that stronger policies generally led to higher costs, but the pattern varied across the locality types.

Application of Retailer caps: The effect on density for retailer caps was straightforward; a 50% cap resulting in half the baseline retailer density. Cost increases were small at less than 3% for both area types at the highest intensities. Urban poor exhibited a slightly greater cost increase (2.78%) at the maximum intensity of 50% compared to the rich environment (1.52%). The cost increase for 90% and 80% caps in urban rich are not statistically significant.

Application of Buffers. Retailer buffer policies predicted a greater impact on costs and density than all other single policies. For urban rich, the predicted difference in cost for a 1500ft retailer buffer was 5.72%. School buffer policies exhibited similar effects in

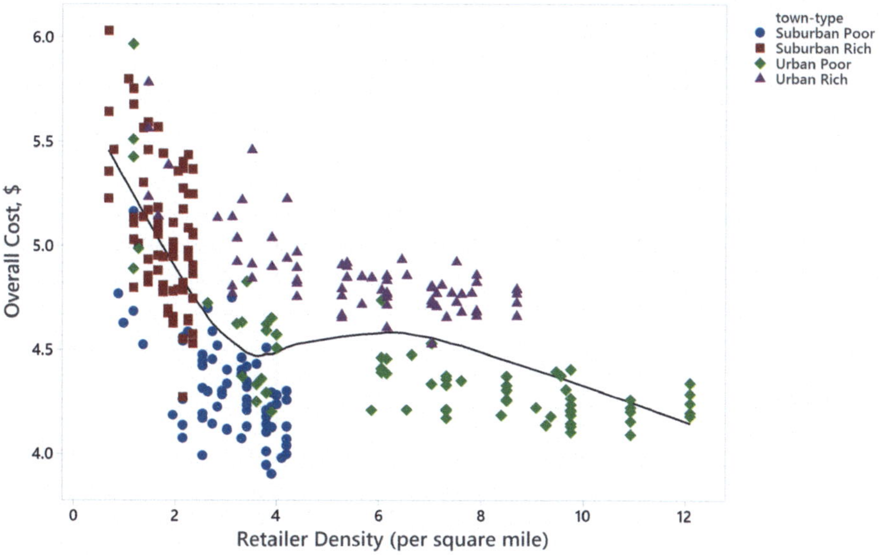

Fig. 3. Retailer Density and Overall Purchase Cost Relationship with Smoothing Spline.

both areas, however the 1500ft policy had a significantly stronger effect in urban poor (2.42%) than in urban rich (1.54%).

Policy combinations: The introduction of a combination of multiple policies predicted the largest impact on overall cost and retailer density. Interestingly the high-intensity combination had a greater effect in Urban poor than in urban rich at 7.63 and 6.42% respectively. The moderate policy had a very similar effect on cost across both areas, notably the effect of the moderate policy was similar to the 1000ft retailer buffer for both areas.

4 Discussion

It was demonstrated that agent-based modelling is a highly useful tool for predicting the effects of public health policies before they are implemented in the real world. The results from the comparative analysis (Fig. 3) highlight that this model is a very accurate representation of the previously created Tobacco Town Agent-Based Model (see graph shown in Fig. 1 in [27]). The contextual effect of policies is considered in this study and the relationship between cost and retailer density is known to exhibit a threshold effect. Cost increases steadily as density decreases up to a point at which costs increase more dramatically—as highlighted in Fig. 3—demonstrating that real policy effects are likely to vary depending on the baseline density of retailers.

For the Glasgow based models, single policies were predicted to exhibit relatively mild effects on cost. However, the model suggested that a combination of policies might be necessary to increase the cost of purchasing cigarettes to a degree that might alter smoking behaviors and reduce the prevalence of smoking in the population.

It is important to consider that the results and data from computational modelling are not usually fully representative of the true situation, however they can act as predictors and be the closest approximation of what would happen when implementing new policies in a real environment [21, 26].

The model reported lower costs on average compared to the original tobacco town model [27]. It was anticipated that the costs reported would be lower than the previous model due to the alternative pathfinding mechanism. The use of updated empirical data may also have contributed to the difference in baseline price. These differences in the models do not affect the mechanics of the model itself and more importantly should not interfere with the effect that retailer reduction policies have on cost.

5 Outlook

Smoking behaviors are constantly evolving, particularly in Scotland. The number of active smokers is steadily decreasing with the introduction of new policy and effective cessation programs. However, with the introduction of electronic nicotine vaporizers or "vapes" there is a new cohort of people purchasing and consuming tobacco free nicotine products in a very similar way to traditional tobacco use. There may be an impetus to apply computational modelling to test policies surrounding the sale of such products, especially since the introduction of mandatory retailer registration.

As the next step in the project, we intend to make Tobacco Town fully data-driven; spatially as well as demographically. For real spatial representation, we can use the GIS data. Figure 4 shows the GIS map of Kelvin Constituency drawn using census and GIS data. For the GIS, the main spatial features extracted from the shape file were roads, built structures, and other non-built structures (like water and open areas). In the model the roads are highlighted in yellow; built structures can be identified as houses (red shades), commercial places (blue), or schools (green). Different shades of house buildings' color represent their accommodation capacity (size and height). The population living in the houses (small dots) was generated from SIMD Data 2020. This can be associated with survey data like the Understanding Society data [31] which can provide each individual its demographic features and other interesting variables (like being a smoker, number of cigarettes a day, income etc.).

Because the Understanding Society's data does not provide exact spatial locations due to privacy reasons, we assigned each smoker a random home and workplace location (certainly based on the accommodation capacity of buildings). Therefore, unless we use another compatible data source, it is not straightforward to validate the spatial characteristics of smokers. However, the Understanding Society provides individual (home) locations at the Lower Layer Super Output Area (LSOA) level which was used as a closed approximation. The retailers (target shapes) were randomly lined along the streets. This is an easier problem to resolve as we can align retailers' numbers and locations using open GIS data (like OpenStreetMap).

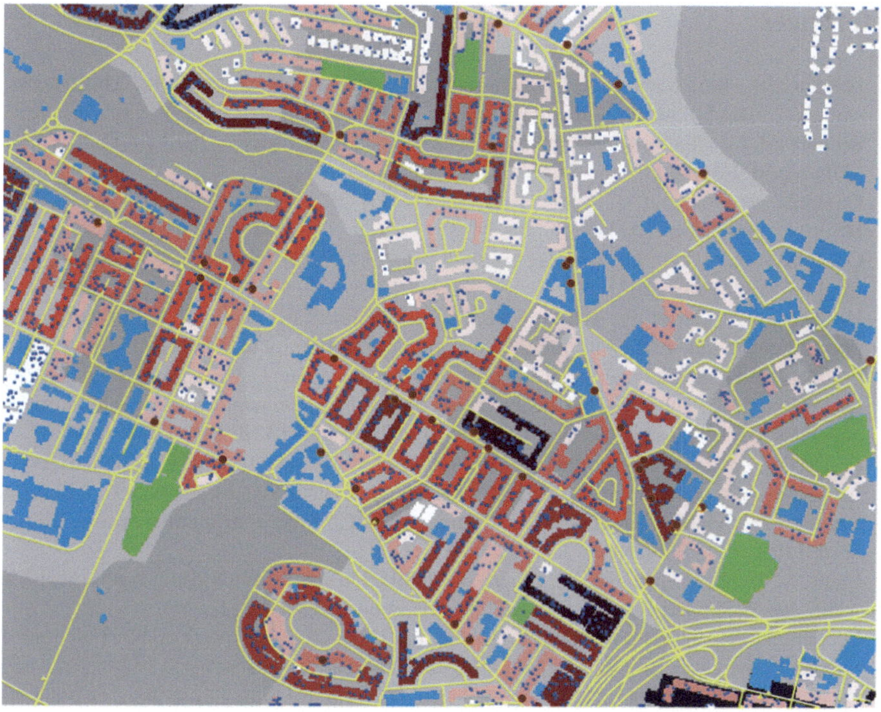

Fig. 4. NetLogo Model of Glasgow Kelvin Constituency

6 Conclusion

In conclusion, this study successfully reproduced The Tobacco Town agent-based model in NetLogo to investigate the effects of tobacco retailer reduction policies on the cost of purchasing cigarettes in Glasgow. The results showed that as retailer density decreases, overall costs increase, and a threshold effect is observed at approximately three retailers per square mile. The study examined the effects of various policy interventions such as retailer caps and buffers, as well as combinations of policies. The results showed that a combination of policies may be necessary to increase the cost of purchasing cigarettes to a degree that might alter smoking behavior. The study acknowledges the limitations of the model due to the time-consuming nature of developing an ABM and the assumptions made due to missing or inaccurate initialization data. The study also exemplifies the capability of multi-agent programmable modelling environments, such as NetLogo, to reproduce and further advance established ABMs. Overall, this study demonstrates the utility of ABMs in predicting the effects of public health policies and highlights the importance of considering the contextual effect of policies on smoking behaviors.

References

1. Allender, S., Balakrishnan, R., Scarborough, P., Webster, P., Rayner, M.: The burden of smoking-related ill health in the uk. Tob. Control. **18**(4), 262–267 (2009)
2. Rutherford, L., Hinchliffe, S., Sharp, C.: The scottish health survey, 2012 edition, volume 1: Main report. Edinburgh: The Scottish Government (2013)
3. Organization, W.H., et al.: World No Tobacco Day 2023: grow food, not tobacco. World Health Organization (2023)
4. Organization, W.H., et al.: Who report on the global tobacco epidemic, 2023: protect people from tobacco smoke (2023)
5. Hollands, G.J., Carter, P., Anwer, S., King, S.E., Jebb, S.A., Ogilvie, D., Shemilt, I., Higgins, J.P., Marteau, T.M.: Altering the availability or proximity of food, alcohol, and tobacco products to change their selection and consumption. Cochrane Database Syst. Rev. (9) (2019)
6. Benninger, M.S.: The impact of cigarette smoking and environmental tobacco smoke on nasal and sinus disease: a review of the literature. Am. J. Rhinol. **13**(6), 435–438 (1999)
7. Pillarisetti, A., Ye, W., Chowdhury, S.: Indoor air pollution and health: bridging perspectives from developing and developed countries. Annu. Rev. Environ. Resour. **47**, 197–229 (2022)
8. Seguel, J.M., Merrill, R., Seguel, D., Campagna, A.C.: Indoor air quality. Am. J. Lifestyle Med. **11**(4), 284–295 (2017)
9. Miall, N., Fergie, G., Pearce, A.: Health inequalities in scotland: trends in deaths, health and wellbeing, health behaviours, and health services since 2000 (2022)
10. Leeming, G., Angus, C., Gillespie, D.: Alcohol and Tobacco Attributable Fractions for Scotland. The University of Sheffield. (2023)
11. Laird, Y., Myers, F., Reid, G., McAteer, J.: Tobacco control policy in scotland: a qualitative study of expert views on successes, challenges and future actions. Int. J. Environ. Res. Public Health **16**(15), 2659 (2019)
12. Scotland, A.: Ash scotland's literature review for nhs greater glasgow and clyde's tobacco planning and implementation group-review of current evidence: Young people and e-cigarettes (2023)
13. Riches, E., Whitehead, R., Rennick, L., Armour, G., McAteer, J., Laird, Y., Reid, G.: What is the causal link between tobacco outlet density and smoking prevalence (2018)
14. Glasser, A.M., Roberts, M.E.: Retailer density reduction approaches to tobacco control: a review. Health Place **67**, 102342 (2021)
15. Berg-Beckhoff, G., Seid, A.K., Stock, C., Bloomfield, K.: Is there an association between home-tobacco outlet proximity and smoking status in denmark? Gabriele berg-beckhoff. Eur. J. Public Health **27**(suppl 3), ckx186–093 (2017)
16. Shortt, N., Tisch, C., Pearce, J., Richardson, E., Mitchell, R.: The density of tobacco retailers in home and school environments and relationship with adolescent smoking behaviours in scotland. Tob. Control. **25**(1), 75–82 (2016)
17. Barnes, R., Foster, S.A., Pereira, G., Villanueva, K., Wood, L.: Is neighbourhood access to tobacco outlets related to smoking behaviour and tobacco-related health outcomes and hospital admissions? Prev. Med. **88**, 218–223 (2016)
18. Caryl, F.M., Pearce, J., Reid, G., Mitchell, R., Shortt, N.K.: Simulating the density reduction and equity impact of potential tobacco retail control policies. Tob. Control. **30**(e2), e138–e143 (2021)
19. Amri, M., Enright, T., O'Campo, P., Di Ruggiero, E., Siddiqi, A., Bump, J.B.: Health promotion, the social determinants of health, and urban health: what does a critical discourse analysis of world health organization texts reveal about health equity? BMC Global Public Health **1**(1), 25 (2023)

20. Maglio, P.P., Sepulveda, M.J., Mabry, P.L.: Mainstreaming modeling and simulation to accelerate public health innovation. Am. J. Public Health **104**(7), 1181–1186 (2014)
21. Tracy, M., Cerd´a, M., Keyes, K.M.: Agent-based modeling in public health: current applications and future directions. Ann. Rev. Public Health **39**, 77–94 (2018)
22. Guzy, M.R., Smith, C.L., Bolte, J.P., Hulse, D.W., Gregory, S.V.: Policy research using agent-based modeling to assess future impacts of urban expansion into farmlands and forests. Ecol. Soc. **13**(1) (2008)
23. Lee, B.Y., et al.: A computer simulation of vaccine prioritization, allocation, and rationing during the 2009 h1n1 influenza pandemic. Vaccine **28**(31), 4875–4879 (2010)
24. Silverman, E., et al.: Situating agent-based modelling in population health research. Emerg. Themes Epidemiol. **18**, 1–15 (2021)
25. Wallace, R.B., Geller, A.B., Ogawa, V.A., et al.: Assessing the Use of Agent-Based Models for Tobacco Regulation. National Academies Press Washington, DC (2015)
26. Lempert, R.: Agent-based modeling as organizational and public policy simulators. Proc. Natl. Acad. Sci. **99**(suppl 3), 7195–7196 (2002)
27. Luke, D.A., et al.: Tobacco town: computational modeling of policy options to reduce tobacco retailer density. Am. J. Public Health **107**(5), 740–746 (2017)
28. Combs, T.B., et al.: Modelling the impact of menthol sales restrictions and retailer density reduction policies: insights from tobacco town minnesota. Tob. Control. **29**(5), 502–509 (2020)
29. Tisue, S., Wilensky, U.: Netlogo: A simple environment for modeling complexity. In: International Conference on Complex Systems, vol. 21, pp. 16–21. Citeseer (2004)
30. Buck, N., McFall, S.: Understanding society: design overview. Longitudinal Life Course Stud. **3**(1), 5–17 (2012)
31. Shortt, N.K., et al.: Using point-of-sale data to examine tobacco pricing across neighbourhoods in Scotland. Tob. Control.. Control. **30**(2), 168–76 (2021)

Author Index

G. Huang et al. (Eds.): ICESD 2024, ESE, p. 189, 2025.
https://doi.org/10.1007/978-3-031-88683-6

The manufacturer's authorised representative in the EU is Springer Nature Customer Service Centre GmbH, Europaplatz 3, 69115 Heidelberg, Germany. If you have any concerns regarding our products, please contact ProductSafety@springernature.com

Printed and bound by CPI Group (UK) Ltd, Croydon, CR0 4YY
01/10/2026
02225936-0001